Astronomers' Universe

Series Editor
Martin Beech, Campion College
The University of Regina
Regina, SK, Canada

The Astronomers' Universe series attracts scientifically curious readers with a passion for astronomy and its related fields. In this series, you will venture beyond the basics to gain a deeper understanding of the cosmos—all from the comfort of your chair.

Our books cover any and all topics related to the scientific study of the Universe and our place in it, exploring discoveries and theories in areas ranging from cosmology and astrophysics to planetary science and astrobiology.

This series bridges the gap between very basic popular science books and higher-level textbooks, providing rigorous, yet digestible forays for the intrepid lay reader. It goes beyond a beginner's level, introducing you to more complex concepts that will expand your knowledge of the cosmos. The books are written in a didactic and descriptive style, including basic mathematics where necessary.

Steven Gullberg • Milton Rojas Gamarra

Inca Cosmovision

The Astronomical Legacy of an Andean Empire

Steven Gullberg
College of Professional and Continuing Studies
University of Oklahoma
Norman, OK, USA

Milton Rojas Gamarra
Universidad Nacional de San Antonio Abad del Cusco
Cusco, Peru

ISSN 1614-659X ISSN 2197-6651 (electronic)
Astronomers' Universe
ISBN 978-3-031-67579-9 ISBN 978-3-031-67580-5 (eBook)
https://doi.org/10.1007/978-3-031-67580-5

© The Editor(s) (if applicable) and The Author(s), under exclusive license to Springer Nature Switzerland AG 2024, corrected publication 2025

This work is subject to copyright. All rights are solely and exclusively licensed by the Publisher, whether the whole or part of the material is concerned, specifically the rights of translation, reprinting, reuse of illustrations, recitation, broadcasting, reproduction on microfilms or in any other physical way, and transmission or information storage and retrieval, electronic adaptation, computer software, or by similar or dissimilar methodology now known or hereafter developed.

The use of general descriptive names, registered names, trademarks, service marks, etc. in this publication does not imply, even in the absence of a specific statement, that such names are exempt from the relevant protective laws and regulations and therefore free for general use.

The publisher, the authors and the editors are safe to assume that the advice and information in this book are believed to be true and accurate at the date of publication. Neither the publisher nor the authors or the editors give a warranty, expressed or implied, with respect to the material contained herein or for any errors or omissions that may have been made. The publisher remains neutral with regard to jurisdictional claims in published maps and institutional affiliations.

Cover image credit: Pinkuylluna Urqu Inti Iluqsimuy (Pinkuylluna Mountain Sunrise). Watercolor printed with permission. © 2024, Jessica Gullberg. All rights reserved.

This Springer imprint is published by the registered company Springer Nature Switzerland AG
The registered company address is: Gewerbestrasse 11, 6330 Cham, Switzerland

If disposing of this product, please recycle the paper.

*We dedicate this book to our loving wives and children, who without their great
support we could not have performed our research and writing.
Alphabetically:
Gregory Richard Gullberg (son)
Jessica Gullberg (wife)
Steven Roland Gullberg II (son)
Sandra Andrea Ortega Olave (wife)
Amaru Phawaq Rojas Ortega (son)
Antarky Rojas Ortega (son)
Ayni Wayra Kawsaypacha Rojas Ortega (daughter)
Gabriel Inti Alejandro Rojas Ferrada (son)
We also dedicate our book to the Native Peoples throughout the Tawantinsuyo.*

Preface

The Inkas worshipped the Sun, and their emperor was thought to be the son of the Sun. (Inka is the Quechua spelling of Inca and will be used throughout the book.) They spread across most of the Andes and evidence of their astronomy exists throughout their former empire. They used solar positions on the horizon for calendrical purposes and managed their crops and religious festivals in this manner. Many examples remain of the intentional light and shadow effects that demonstrate their sophisticated understanding of the Sun's movement and of solar horizon events.

Their astronomy can only be fully understood in its cultural context, and that is the focus of this book. *Inca Cosmovision: The Astronomical Legacy of an Andean Empire* explores the cosmic worldview of the Inkas from the perspective of oral traditions passed from one generation to the next among the Inkas' living descendants. You will learn about Inka astronomy in a way that you perhaps have never encountered. Milton Rojas Gamarra, one of the book's two authors, is a Quechua descendant of the Inkas. What you will read benefits from his and author Steven Gullberg's field research, but most significant is what you will experience from what we have written from the many stories Milton learned from his grandparents and great-grandparents, and from his *Amauta*, the term for a highly respected Indigenous teacher of Inka culture. Quechua is Milton's first language, so you will also benefit from the insight provided through the use of Quechua words. Doing so with the native language is important for a better understanding.

The chapters in the book that are based upon Milton's oral traditions must be viewed as such. There are no scientific references to cite. Instead, these are beliefs that have existed since the time of the Inkas, and in the minds of many Quechua descendants they still do. By embracing what is shared you will be learning about the Inkas and their astronomy from their perspective, in such a way that has not previously been available. As you read you should immerse yourself in what is presented and enjoy this from the Inka worldview rather than the worldview of Western science or worldviews from previously attempted interpretations.

It also is our goal to capture these ideas from oral traditions and put them into print before they are lost. This book will enlighten you about Inka cosmovision to a depth that no other has before. We hope that you enjoy the journey!

Norman, OK, USA Steven R. Gullberg
Cusco, Peru Milton Rojas Gamarra
May 2024

Acknowledgments

We would like to thank author Milton's Amauta Emilio Huaman Huillca because without his insight this book would not have been possible. We would also like to acknowledge Milton's grandparents and great grandparents for their sharing of oral traditions with him. We as well acknowledge all those who were interviewed for the sharing of their knowledge regarding oral tradition and Inka astronomy and cosmovision.

Additionally, we most certainly must give our thanks to Jessica Gullberg for the wonderful watercolor paintings that she created to help us illustrate.

Contents

1	**Introduction**	1
2	**Inka Chronology**	5
	Early Andeans	7
	Early Inkas	8
	The Period of Expansion	9
	Pachakuteq Yupanqui Inka	10
	Topa Inka Yupanqui	11
	Wayna Qhapaq	12
	The Spanish Conquest	12
	The Catholic Purge	13
	Summary	14
3	**Facets of Inka Culture**	15
	Religion	16
	Cosmology	18
	Landscape	19
	Camay	19
	Sacred Animals	20
	Ancestors	20
	Royal Marriage	21
	Moieties, Suyus, Panaqas, and Ayllus	21
	Money	22
	Records	22
	Khypus	23

	Textiles	23
	Festivals	23
	Climate	26
	Agriculture	26
	Irrigation	27
	Pilgrimage	27
	Architecture	28
	Qhapaq Ñan	34
	Carved Rocks	36
	Summary	37
4	**Inka Culture and Quechua Language**	**39**
	Time, Space, and Language	42
	Kawasaypacha	43
	Ayni	44
	Tawantinsuyo	45
	Inka	48
	Ayllu	49
	Panaqa	49
	Waqa	49
	Apu	50
	Apus and Kawsaypacha	50
	Apus, Waqas, and Ayni	50
	Ushnu	51
	Minka and Mit'a	51
	Sumaq and Allin	52
	Tinkuy	53
	Iskaynintin	54
	Masintin and Yanantin	54
	Mass	55
	Yana	55
	Llaqta	56
	Paqarina	56
	Inka Creation	56
	Usufruct	57
	Currency	58
	Summary	58

5 Archaeoastronomy — 59
- Celestial Sphere — 59
- Heavenly Motions — 60
- Mid-latitudes — 60
- Qosqo — 61
- Solstices — 61
- Equinoxes — 61
- Zenith and Anti-zenith Sun — 62
- Horizon Astronomy — 63
- Cultural Astronomy Tools — 63
- Horizon Deviation — 64
- Summary — 66

6 Facets of Inka Astronomy and Cosmology — 69
- Qorikancha — 70
- Solar Worship — 70
- Cosmological Origins — 71
- Festivals — 71
- Alignments in Architecture — 72
- Ushnus — 73
- Sukanqas — 73
- Pillars — 75
- The Milky Way — 75
- Orientation and Quadripartition — 75
- Celestial River — 76
- Dark Constellations — 77
- Star Constellations — 77
- Stars — 77
- Planets — 78
- The Pleiades — 78
- Seq'e System and the Stars — 79
- Q'enqo Grande — 79
- Lacco — 82
- Waqa 44 — 84
- Q'espiwanka — 85
- Machu Piqchu Region — 86
- Summary — 92

7	**The Waqas of the Inkas**	97
	Summary	126
8	**Inka Seq'es**	129
	Seq'es	130
	The Qosqo Seq'es Emanated from the Qorikancha	133
	The First Three Seq'es of Chinchaysuyo and their Waqas	136
	Qollao	137
	Michosamaro	137
	Patallaqta	139
	Pillcopuquio	139
	Cirocaya	140
	Sonconancay	141
	Payan	142
	Guaracince	142
	Racramirpay	143
	Intiillapa	144
	Viroypacha	144
	Chuquibamba	145
	Macasayba	145
	Guayrangallay	146
	Guayllaurcaja	146
	Collana	147
	Nina	147
	Canchapacha	148
	Ticicocha	149
	Summary	149
9	**Inka Constellations**	151
	Types of Inka Constellations	153
	Principal Inka Constellations	153
	Qollca	154
	Urquchillay and Qatachillay	155
	Dark Constellations	155
	Ch'aska Punchu, Chacra, Orqorara, Chakana, Quntur, Suyuntuy, and Huaman	159
	Chuchuqoyllor	160
	Mallki	161
	Kotu Sankha	161

	Puma Yunta	161
	Laja Haykuna	161
	Summary	161

10 Inka Astronomy in Daily Life — 163
　Daily Life and Routine — 164
　Everyday Knowledge — 164
　Astronomy in the Daily Life of the Inkas — 165
　Examples of Astronomical Ceremonies and Participation
　of the People — 165
　Example of Archaeological Places and Frequency of Visits
　of the Town — 169
　Summary — 171

11 Conclusion — 173
　Final Thoughts — 174

Correction to: Inca Cosmovision: The Astronomical Legacy of an Andean Empire — C1

Bibliography — 177

Index — 183

About the Authors

Steven R. Gullberg holds a Ph.D. in astronomy from James Cook University (Australia) and is a Professor of Cultural Astronomy at the University of Oklahoma (USA), where he is Lead Faculty for the School of Integrative and Cultural Studies. He additionally serves as President of the International Astronomical Union (IAU) Commission C5 for Cultural Astronomy, and as well as the managing editor of the *Journal of Astronomy in Culture*, the journal of the International Society for Archaeoastronomy and Astronomy in Culture (ISAAC). He has conducted extensive field research on the astronomy of the Inkas in the Peruvian Andes and has published many research papers. He also is the author of *Astronomy of the Inca Empire: Use and Significance of the Sun and the Night Sky*. At the University of Oklahoma, he led the development of archaeoastronomy distance-learning courses designed to educate researchers around the world. Dr. Gullberg is regularly invited to give talks at international conferences as he endeavors to globally advance the field of Cultural Astronomy.

Milton Rojas Gamarra earned his Ph.D. in Intercultural Education at the University of Santiago (Chile). His doctoral thesis was in archaeoastronomy and his master's degree was in astrophysics. Milton is a Professor of Physics at the Universidad Nacional de San Antonio Abad del Cusco (Peru). Milton is a descendant of the original Quechua people and has dedicated himself to preserving their wisdom, their ethos, and their worldview. His first language is Quechua. Dr. Rojas Gamarra has studied the astronomy of the Inkas for more than 30 years and benefits greatly from the oral traditions that have been passed on to him by his grandparents and by his *Amauta*, Emilio Huaman Huillca. Milton shares his wonderful insight in this book.

1. Introduction

Civilizations create their own culture, their own ethos, and their own worldview, and this was certainly true of the Inkas. The Inka Empire (in Quechua, the Tawantinsuyo) in its evolution developed its own customs, habits, ways of being, and ways of behavior. The Inkas as well sought to answer fundamental questions such as "how to be" (*Imaynakay*), and "how to live" (*Imaynakawsay*) (Fig. 1.1).

Like other ancient civilizations, the Inkas tried to interpret and understand the cosmos—its origins, its primordial stages, and to seek a sense of its existence and through this they developed their worldview (*Kawsaypacha*). In the process, the Inkas built their own principles of life (*Kawsay*), standing out among them gratitude and reciprocity (*Ayni*—the force of reciprocity and help) and the creative life force of the *Pachamama*, *Kawsaypacha*, which means everything in the cosmos lives. Ayni is a reflection of the reality that exists in the energy of the world.

Kawsaypacha corresponds to energy in time and space. This book addresses these principles and describes how Andean culture was based on a rationality different from that of the European West—the principles that govern Andean thought are based on transversal concepts and principles of rationality, integrality, and cyclicality. Thus, a proper understanding and interpretation of the Inka legacy in cosmovision and astronomy can only be carried out properly when viewed within the context of Inka culture and its worldview.

The chapters that follow begin with the history of the amazing Inka Empire, its construction, and its expansion by conquest until its own conquest by the Spaniards. We will describe facets of Inka culture and you will learn about Quechua, the language of the Inkas, the language still predominant

Fig. 1.1 Traditional Quechua celebration dress, author Milton second from the right. (Watercolor printed with permission. © 2024, Jessica Gullberg. All rights reserved)

throughout rural areas away from cities such as Lima and Qosqo (Cusco). A general understanding of the nature of Quechua is essential for truly sensing cultural context and after words in Quechua are introduced, we will continue to use them in order to give you a better sense. The book will introduce you to fundamentals of Archaeoastronomy and then will continue with facets of Inka astronomy. *Seq'es* and *wakas* and their relationships to Inka astronomy are discussed in detail before going on to explore constellations of the Inkas. We will examine Inka worldview and culture, including their perception of space and time and show how this all existed in daily life.

The Inkas organized their civilization with seq'es and waqas and both figure prominently in their cosmology. Waqas can be roughly thought of as shrines that often were believed to be sentient with great power and influence. The waqas needed to be cared for and thus were located on seq'es which were organizational lines, not always straight, with responsibilities for their care assigned to extended family lineages. There were 41 seq'es surrounding Qosqo and their astronomy, cosmology, and calendrical functions are explored.

Archaeoastronomy is the study of how astronomy was used by ancient cultures. Fundamentally important is placing what is found into cultural context because without doing so the data collected can be easily misunderstood.

Having set the stage, you will learn more specifically about the astronomy that was used in Inka culture and the culture's timekeeping. This will be a fascinating blend of astronomical and archaeoastronomical knowledge derived greatly from insight that has been passed from one Quechua generation to generation via oral traditions from the time of the Inkas up to today. The Inkas worshipped the Sun and thought their emperor to be the son of the Sun and his wife to be the daughter of the Moon.

Machu Piqchu may only be reached by train or by hiking. The Inka Trail (Qapaq Ñan) is the most popular route and the trek from Ollantaytambo lasts at least 3–4 days. Fascinating examples of Inka astronomy and cosmovision can be found along the way and many more are present at the journey's end in Machu Piqchu and Llaqtapata. Examples of intentional light and shadow effects and solar orientations abound throughout the empire and photographic examples from field research are shared. Original artwork enhances the book's illustration as well. Along with the Quechua cultural insight provided, these images will help to give you a visual sense of the astronomy of the Inka empire and its cosmovision.

Elements of Inka culture converged with science and astronomy, but the worldview of the Inkas was conceptually different from that found in what is referred to as Western culture. Not only does Inka astronomy have a different orientation, but the empire was filled with many sacred entities with great powers. This book, through the approach of employing Quechua terms and actual oral traditions passed down to author, Milton, will enable you to gain a much better understanding and appreciation of the cosmovision of the Inkas.

2

Inka Chronology

Contents

Early Andeans	7
Early Inkas	8
The Period of Expansion	9
Pachakuteq Yupanqui Inka	10
Topa Inka Yupanqui	11
Wayna Qhapaq	12
The Spanish Conquest	12
The Catholic Purge	13
Summary	14

An important part of understanding Inka culture is to know their history. By the time of the sixteenth century invasion from Europe the Inka empire had grown across the Coordillera de los Andes (Andes Mountains) and coastal regions from Argentina and Chile through Bolivia and Peru to Ecuador and into southern Columbia (see Fig. 2.1). The period of conquest for the Inkas was relatively short, but they learned from the many civilizations that preceded them. Celestial myths, beliefs, and astronomical knowledge developed with observations made over the centuries; the Inkas inherited this knowledge and adapted it to their needs. They instituted a state solar religion founded upon what they had learned, and temples and shrines were built to support this with displays of light and shadow (Gullberg, 2020).

Further advancement essentially came to an end by 1532 CE when Spanish conquistadors invaded Peru. The Spaniards did not appreciate local astronomical knowledge and instead continued with the European traditions. Catholic priests searched to locate and destroy everything related to Inka religion, including anything they could find related to worship of the Sun, Moon, and the stars (Gullberg, 2020).

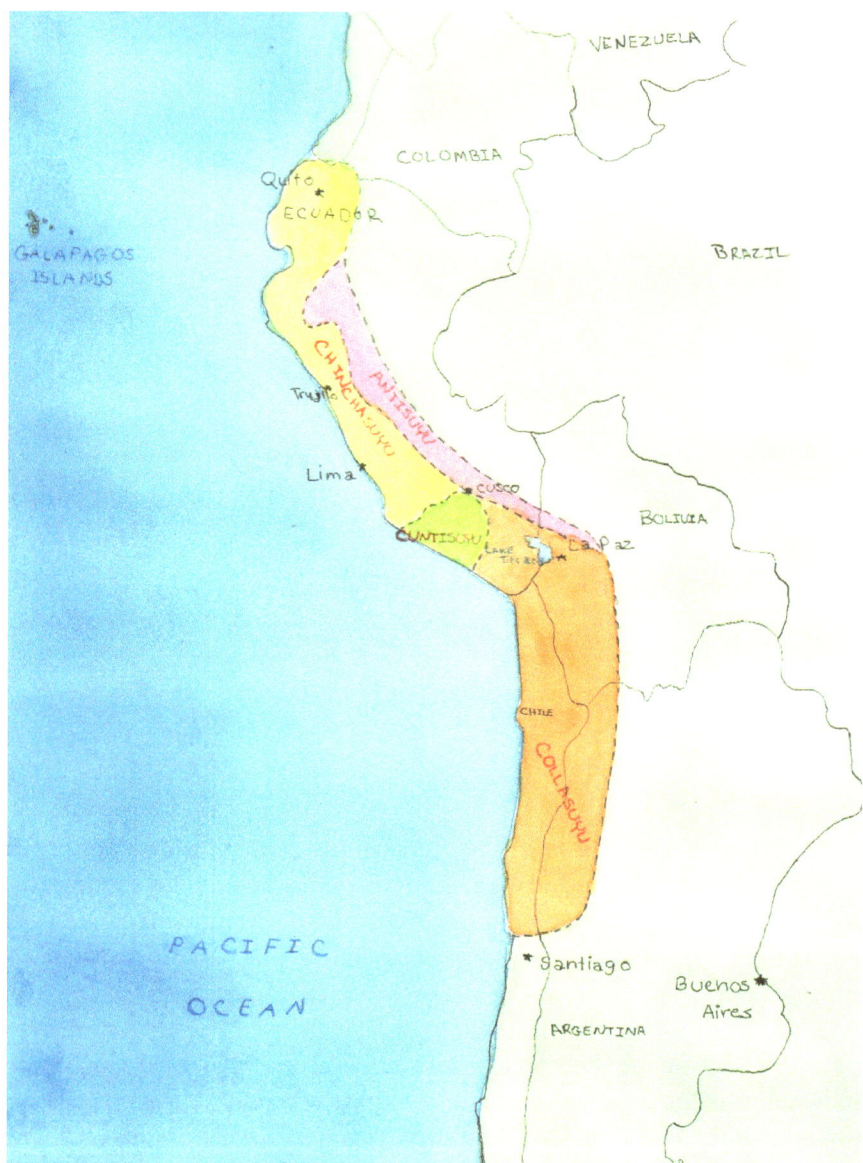

Fig. 2.1 The Tiwantinsuyu (Inka Empire) and its four suyus. (Watercolor by Jessica Gullberg modified with permission from (Staller, 2008). © 2008, Springer Science+Business Media, LLC. All rights reserved)

Early Andeans

The civilization of the Inkas was built upon what was learned from cultures that preceded them, such as the Chavín, Moche, and Huari. In the Middle Formative much of Peru was influenced by the Chavín and what we know of them is primarily from their art (Burger, 1992). During the Late Formative a culture well north of Lima and not far from the Pacific coast of which we know very little built what is now called the Chankillo Archaeoastronomical Complex (UNESCO, 2021). There 13 massive towers, when viewed from a distant observing point, have calendrical implications with the June solstice Sun rising at the left extreme and the December solstice Sun rising at the rightmost tower (see Fig. 2.2). In the Early Intermediate, to the north lived the Moche, known for their pottery and textiles. The Huari followed in the Middle Intermediate and Middle Horizon, and later the Chimú in the Late Intermediate. In the region of southern Peru in both the Early and Middle

Fig. 2.2 June solstice sunrise by the left-most tower of the Chankillo Astronomical Complex. (Watercolor printed with permission. © 2024, Jessica Gullberg. All rights reserved)

Intermediate were the Nasca who produced pottery and textiles as well as enigmatic lines thought to be astronomically related (Gullberg, 2020).

In the Middle Horizon much of Peru was dominated by the Huari and the Tiwanaku. The Huari lived in Peru's Ayacucho region while the Tiwanaku were located near Lake Titicaca in Bolivia. The Inkas, with their cosmological beginnings at Lake Titicaca, likely learned much from the Tiwanaku (Hemming, 1970; Gullberg, 2020).

During the Late Intermediate other cultures arose, such as the Canarí, Chanca, Colla, Lupaca, Huanca, Conchuco, Yarivilca, Chachapoya, and the Inkas. The other tribes thrived until they each were assimilated by the Inkas (Hemming, 1970). Monumental sculpture of the Inkas seems likely to have originated with what was learned from the Tiwanaku and Huari. And Tiwanaku cosmology helped to provide the Inkas with the astronomical concepts that they needed to support the legitimacy of their empire (Zuidema, 1977; Hemming & Ranney, 1982; Gullberg, 2020).

Early Inkas

It is thought that the earliest Inkas may have settled in the Qosqo (Cusco) valley circa 1200 CE. Inka mythology describes that Manqo Qhapaq (Manco Capac) was the first Inka ruler, having descended from the god Wiraqocha (Salazar & Salazar, 2014). Manqo Qhapaq began the dynastic succession that led to the eighth emperor, Wiraqocha Inka, and his son, Inka Yupanqui, or Pachakuteq (Pachacuti). Niles says that there is no evidence for mummies of the first four Inkas and suggests they could be mythical (D'Altroy, 2002; Niles, 1987). The royal lineage between Manqo Qhapaq and Wiraqocha Inka was Sinchi Roca, Lloque Yupanqui, Mayta Qhapaq, Qhapaq Yupanqui, Inka Roca, and Huahar Huacac (Gullberg, 2020).

The Inkas asserted that they were the chosen people of the Sun and that they had been "created to be the rulers" (Sherbondy, 1992). "They were the children of the sun and therefore first among all peoples" (Sherbondy, 1992, pp. 55–56). A pivotal occurrence took place c. 1438 when the Chancas decided to attack the Inkas. Wiraqocha Inka, sensing defeat, fled Qosqo with his first son, leaving defense to a younger brother, Inka Yupanqui. In the ensuing battle he routed the Chancas and after the victory he usurped rule from both his father and brother (D'Altroy, 2002; Zuidema, 1964). He adopted the name Pachakuteq Yupanqui Inka and set out to conquer the allies of the

Chancas. He did not stop there, however, and continued to expand Inka rule over all known tribes throughout the Andes (Gullberg, 2020).

Pachakuteq began a great building program that created a majority of the waqas, palaces, temples, and *Qhapaq Ñan* (Inka roads) that can still be seen today. Tribes, such as the Chimú, when defeated were often dispersed throughout the Empire as workers called *mitmaes,* this to diffuse threats to Inka security and also to take advantage of specialists with skills needed by the state (D'Altroy, 2002; Gullberg, 2020).

It is thought that the Tiwanaku might have been an inspiration for the Inka's monumental sculpture (D'Altroy, 2002; Paternosto, 1989). There is a gap between the two cultures, but similarities do exist in their architectural styles. Additionally, knowledge of astronomical cycles can only be learned over long periods of time so, in a similar fashion, the Inkas must have acquired their knowledge of astronomy from earlier civilizations (Gullberg, 2020).

Pachakuteq, his son, and grandson respectively built the largest empire ever known in the pre-Columbian Americas, 4800 km from present-day Chile to Columbia (see Fig. 2.1). Qhapaq Ñan (roads) were built to suppport armies at distant stations, as were Qollqa (storehouses). Temples and shrines with solar orientations were constructed to support the state control of subjects. Pilgrimage centers were also built to reinforce the legitimacy of royal rule (Bauer and Stanish 2001; Gullberg, 2020).

By the time Francisco Pizarro arrived in Peru in 1532 the Inkas had not only built an impressive empire, but also a remarkable society. Their civilization had advanced to its sophisticated level without European influence or even knowledge that nations such as Spain existed (Hemming, 1970). As far as "the son of the Sun" was concerned, he ruled the world (Gullberg, 2020).

The Period of Expansion

The greatest period in Inka history began with Pachakuteq's ascent c. 1438 and lasted until the c. 1527 CE death of his grandson, Wayna Qhapaq. Between them, Pachakuteq's son and Wayna Qhapaq's father, Topa Inka, governed from 1471 to 1493 (Niles, 1987). Pachakuteq ordered massive construction, developed a complex society, created a domineering religion, and displayed a great interest in astral alignments. His son and grandson continued his visions of expansion and construction, albeit with interpretations of their own (Gullberg, 2020).

Pachakuteq Yupanqui Inka

Pachakuteq (Pachacuti), the name adopted by Inka Yupanqui, translates as "transformer of the world." More specifically, in Quechua, *pacha* means "a moment or interval in time and a locus or extension in space" (Salomon & Urioste, 1991, p. 14) and *kuteq* "to turn around" (Hemming and Ranney 1982; Gullberg, 2020; see Fig. 2.3). Pachakuteq saw himself as the son of the Sun and co-creator of the land. Salazar (2004) suggests that Pachakuteq felt a special association with "supernatural forces imminent in the landscape and the celestial sphere," and that his connection with these forces needed to be "actively reaffirmed through daily ritual" (p. 41).

The Inkas first sought to control the Chancas and then set out to subjugate the rest of their known world. Pachakuteq wanted to spread Inka religion and

Fig. 2.3 Pachakuteq Yupanqui Inka statue as part of a fountain in Qosqo. (Watercolor printed with permission. © 2024, Jessica Gullberg. All rights reserved)

culture across the Andes (Niles, 1987) and he sought to assimilate tribes peacefully without bloodshed. The Inkas would send scouts to assess a tribe and contact its chief informing him that he could keep his throne if they willingly joined the empire. It was made clear that the Inkas would take the land and the people by force if necessary. Otherwise, the tribe could be welcomed as new citizens of the empire and, if not, be destroyed by the Inka army (Hemming, 1970; Gullberg, 2020).

Pachakuteq carried on a campaign of conquest across the Andes that was continued during the reign of his successors, Topa Inka and Wayna Qhapaq. These conquests were not always peaceful, though. Subjugation of the Cañari, for instance, was protracted and quite brutal. Cañari loyalty was never secured and they eagerly assited the Spanish conquistadores against their Inka masters (Hemming, 1970; Gullberg, 2020).

Pachakuteq felt he both could and was obliged to improve upon the handiwork of his creator. As the son of the Sun and co-creator of the land he promoted a style of masonry that was blended with natural rock formations. Several sites give the appearance that manmade stone blocks simply grow from their natural rock foundations (Paternosto, 1989). Pachakuteq also enhanced stone outcroppings to improve them as waqas, and some included astronomical alignments. Celestial orientations found in these waqas display a strong interest in the solstices (Gullberg, 2020).

Pachakuteq established a calendar for the timing of religious ceremonies and rituals, and as well for the proper times for the planting and harvesting of specific crops (Bauer and Dearborn 1995). This was based on recurring astral cycles, especially those involving the solstices and the heliacal rise of the Pleiades (Gullberg, 2020).

Topa Inka Yupanqui

Topa Inka acceded to the throne in 1471 CE upon his father's death and continued the territorial expansion. Topa Inka chose land in the Chinchero valley for his rural estate which extended to the salt terraces of Maras. Natural rock and landscape features were utilized along with those that were carved (Niles, 1999). Topa Inka's walls and terraces at Chinchero are oriented cardinally, precisely north-south and east-west. The cardinal direction of south might first have been determined by the shadow plot of a vertical gnomon. North, east, and west would then follow geometrically. Topa Inka is also credited with completing the structure of Saqsaywaman at the northwest edge of Qosqo. He ruled until 1493 (Niles, 1999; Gullberg, 2020).

Wayna Qhapaq

Wayna Qhapaq (Huayna Capac) was the son of Topa Inka and Mama Oqllo who, as his father's wife and sister, was both his mother and his aunt. His parents, therefore, belonged to a lineage that was both patrilineal and matrilineal, just as did the creator god Wiraqocha (Zuidema & Quispe, 1973; Gullberg, 2020).

Wayna Qhapaq was born at Tomebamba, near the present-day Cuenca in Ecuador. Niles (1999) stated that Pachakuteq chose this of his grandsons to succeed Topa Inka in the dynasty. She describes how Wayna Qhapaq was close with his mother and she made him promise not to leave her to do battle, and he kept this pormise until her death (Gullberg, 2020).

Qoya Cusirimay, Wayna Qhapaq's legitimate wife, was his full sister, but she did not give birth to an heir (Niles, 1999). He next chose another sister, but she did not receive the blessing of his father's mummy. Eventually, he took Cibichimpo Rontocay as his principal wife. Wayna Qhapaq fathered many children with many wives, among them were his sons Huascar and Atawallpa. Atawallpa was the elder and they were half-brothers. When Wayna Qhapaq left Qosqo to subdue rebellious tribes in the north, he left Huascar behind to govern and took Atawallpa with him into battle (Niles, 1999). Wayna Qhapaq built his country estate in the north of the Sacred Valley placing his palace, Q'espiwanka, near the present-day village of Urubamba. Q'espiwanka as well displays interest in the solstices with its astronomical alignments (Gullberg, 2020).

Wayna Qhapaq died in 1527 CE without having designated an heir, possibly due to the sweeping epidemic of smallpox from livestock that came with arriving Europeans to the north. His mummy was hidden at Q'espiwanka, but the Spaniards discovered it more than 20 years later (Betanzos, 1996 [1576]; Gullberg, 2020).

The Spanish Conquest

Fifty-four years after Pachakuteq defeated the Chancas, Columbus sailed to the West Indies in 1492 CE. Once word of the riches he found reached the Old World, the new one would never be the same (Hemming, 1970). In 1519 Hernán Cortés invaded Mexico and set out to conquer the Aztecs with as few as 500 men and 16 horses. Cortés was helped by some of the Aztec's subject tribes and succeeded with the conquest; he shipped much wealth to his king

(Hemming, 1970). The promise of such riches lured many to the Americas in search of personal fortunes (Gullberg, 2020).

In 1522 Pascual de Andagoya explored the Columbian coastline in search of the *Viru*, the name which later inspired *Peru*. Francisco Pizarro set out in 1524 and 1526 to find his fortune but failed on both occasions (Cieza de León, 1998 [1555]; Hemming, 1970). The Spaniards in Central America brought diseases such as smallpox, measles, and diphtheria for which the Andeans had little resistance. As these plagues spread, daily routines were abandoned and Inka society faltered (Gullberg, 2020). Meanwhile a civil war between Huascar and Atawallpa ensued for control of the empire. Atawallpa ultimately prevailed and captured Qosqo from Huascar in 1532.

In 1529 Pizarro was authorized to conquer Peru and would be made its governor. He was joined by three of his half-brothers and they sailed from Panama to Ecuador in 1530, ultimately reaching the northern boundaries of the Inka empire in 1531. Soon after Atawallpa conquered Qosqo he was captured by Pizarro. The Spaniards are said to have advanced into Peru with 62 cavalrymen and 106 soldiers, at the time the Inka civil war was coming to an end. With their technological superiority of armored and mounted soldiers, and the lingering effects of plague and the civil war, the Inkas were defeated (Hemming, 1970; Gullberg, 2020).

Pizarro held Atawallpa captive for 8 months before finally executing him in July of 1533. The Spaniards showed little interest in Inka astronomy and at the time of the conquest were still influenced by Claudius Ptolemy's geocentric system. Copernicus' heliocentric theory of the universe was first introduced in Europe in 1543 but it received little acceptance for decades (Gullberg, 2020).

The Catholic Purge

The Catholic Church set out to convert all Inkas to Christianity (Hemming, 1970). Spain requested permission from the Vatican to conquer the Americas and was granted this by Pope Alexander VI, as long as they converted the Indigenous there to become Christians. Chapels, cathedrals, and monasteries were built in the major centers of Peru (Hemming, 1970; Gullberg, 2020; see Fig. 6.1).

In 1539 the Spanish began a ruthless campaign of extirpation against the Indigenous religion and destroyed as many waqas as possible (Arriaga, 1968

[1621]; Bauer, 1998; Gullberg, 2020). In 1567 an ecclesiastical council was conducted in Lima for the eradication of all pagan rites. Hemming (1970) relates that priests were instructed to abolish superstitions, ceremonies, arrest witch doctors, and destroy any shrines or talismans. The Church began an aggressive effort to erase all elements of Indigenous religion. Father Pablo Joseph de Arriaga described in detail the process of idolatry extirpation in Peru. Arriaga (1968 [1621]) said that the "moveable huacas" once discovered were taken away and burned. Huacas that could not be removed, such as "high hills and mountains and huge stones" could still be worshipped. Of these he stated "…we must try to root them out of their hearts, showing them truth and disabusing them of error" (pp. 24–25). Inka temples were either destroyed or their stone bases were used as the foundations for such as Catholic chapels (Gullberg, 2020).

Summary

An in-depth understanding of Inka culture and what influenced its cosmological and religious beliefs is required before understanding why these astronomical orientations grew to be of such great importance. The Inka empire was relatively short-lived, but the Inka culture had been built upon what was learned from the many societies that preceded them. This was critical for Inkan astronomy because celestial knowledge had to be obtained through observations over very long periods of time. Astronomy was a major part of Inka cosmology, religion, and agriculture and the significance of such is evident in many of the extant waqas and structures left by the emperors Pachakuteq, Topa Inka, and Wayna Qhapaq.

Our knowledge of Inka astronomy has been limited due to the lack of a written language and the desire of the Catholic Church to eradicate anything viewed as being related to Indigenous religion. Records of huacas, however, were maintained by the Church after waqas were destroyed, and Spanish chroniclers also recorded information related to them and to astronomy taken in conversations with local informants. The chronicals do not give true contextual perspective to Inka astronomy, however, because the writings were framed from a European point of view at a time when even the teachings of Copernicus were new. The Spaniards failed to realize that the Inkas saw the cosmos in their own unique and sophisticated way. In chapters ahead gaps are filled with what can be learned from Inka oral traditions.

3

Facets of Inka Culture

Contents

Religion	16
Cosmology	18
Landscape	19
Camay	19
Sacred Animals	20
Ancestors	20
Royal Marriage	21
Moieties, Suyus, Panaqas, and Ayllus	21
Money	22
Records	22
Khypus	23
Textiles	23
Festivals	23
Climate	26
Agriculture	26
Irrigation	27
Pilgrimage	27
Architecture	28
Qhapaq Ñan	34
Carved Rocks	36
Summary	37

Astronomy was an integral part of Andean mythology and creation, and it was at the very heart of Inka culture for religion and agriculture. To understand Inka astronomy, it must be placed into context in the greater society. Why was it important, why was it used, and how was it used? This chapter will lay the groundwork for aspects of Inka culture and its association with astronomy.

Fig. 3.1 The Torreon and Wayna Piqchu at Machu Piqchu. (Watercolor printed with permission. © 2024, Jessica Gullberg. All rights reserved)

This then will be built upon in subsequent chapters to better understand Inka thought and reasoning through the use of oral traditions (Fig. 3.1).

Religion

The Inkas proclaimed that they were the children of the Sun. They worshipped the Sun and viewed their emperor as being its direct descendant. They learned from existing Andean astronomical knowledge and beliefs and made solar worship the official religion of their empire. Pachakuteq imposed it across the realm, maintaining that he was the son of the Sun and his wife the daughter of the Moon. Religion was closely tied with nature and with the prosperity of the world brought about by the supernatural forces of mountains, caves, and streams, and as well as those in waqas and celestial objects such as the Moon, stars, rainbows, and thunder (Hemming, 1970; Hemming & Ranney, 1982). Assimilated tribes had to accept Inka religious beliefs but were allowed to

Fig. 3.2 Drawing of an Inka emperor procession; such processions could be for the living or the dead. (Reprinted with permission from (Guaman Poma, 2010). © 2010, University Texas Press. All rights reserved)

continue worship of their lesser gods as well. The ruling Inka was the central figure in solar worship, and this supported that he was the descendant of the Sun. (Bauer and Dearborn 1995; Bauer & Stanish, 2001). This deification established the Inka's legitimacy and justified his absolute authority (Hemming, 1970; Gullberg, 2020).

The Inkas associated *Qoya*, the queen of the ruling Inka, with the Moon (Bauer & Stanish, 2001). The Moon served as both the wife and the sister of the Sun, a relationship that also existed with the empire's ruling couple (Zuidema & Quispe, 1973). The Moon, in Inka culture, was feminine and the Inka's Qoya was its daughter. Women worshipped the Moon and made offerings to it during eclipses or when giving birth. Descriptions of Inka lunar worship are few, however, perhaps because of its feminine role in society (Bauer and Stanish 2001). Worship of the Sun took place at temples and waqas, as well as in pilgrimage centers designed to promote the authority of the ruling elite (Hemming, 1970; Gullberg, 2020; see Fig. 3.2).

Cosmology

The Inkas believed the world was created by their god Wiraqocha at Lake Titicaca. Wiraqocha was the father of the Sun and the Moon. He was both male and female which enabled him to be the founder of both patrilineal and matrilineal descent lineages (Zuidema & Quispe, 1973). Wiraqocha first made people of stone and then made the Sun, stars, and Moon. He gave the stones life as they appeared from caves, rocks, and springs (Paternosto, 1989; Gullberg, 2020).

A common Inka creation myth was that Manqo Qhapaq with his three brothers and four sisters, left Lake Titicaca in a migration beneath the Earth and they emerged from a T'oqo (a cave) south of Qosqo (Hemming and Ranney 1982). This is important because caves and other natural features of the Earth were venerated by the Inkas. They viewed caves and rock outcroppings as connections with their underworld, Ukhupacha (Gullberg, 2020).

Inka cosmology therefore starts at Lake Titicaca, a large lake at an altitude of 3810 masl on the border between present-day Peru and Bolivia. The waters create a microclimate conducive for agriculture, even at that altitude (Bauer and Stanish 2001). Pachakuteq conquered the Lake Titicaca region. Inka legend has it that the Island of the Sun and the Island of the Moon were respectively the points of origin for the Sun and the Moon, and therefore for the entire Inka tribe. Pachakuteq was quick to incorporate these shrines into Inka origin myths (Bauer and Stanish 2001). He constructed a temple of the Sun and shrine to the Moon on the islands and, as the Inkas' cosmological points of origin, they were instituted as state-sponsored ritual pilgrimage centers (Bauer and Stanish 2001; Gullberg, 2020).

The Inkas believed that the Sun first rose from a sacred rock called Titicaca on the north end of the Island of the Sun, and annual rituals were held there at the times of both the June and December solstices (Bauer and Stanish 2001; Gullberg, 2020).

The cosmos of the Inkas existed in three distinct worlds—that of Ukhupacha, the underworld, Kaypacha, the here and now, and Hananpacha, the world above (Urton, 1981). There are many extant examples of sets of three stairs that represent symbolic transition between these three worlds of the Inkas (Gullberg, 2020).

Landscape

The Inkas venerated natural features such as mountains (which were Apus), outcroppings, caves, springs, and rivers, and all were believed to be endowed with sacred powers. They gave high reverence to Apus and the great entities within them. Quechua populations today continue to view Apus as powerful deities or as residences of deities. They are worshipped as ancestors, sources of water and weather, and in the case of Nevado Ausengate, the father of alpacas and llamas. These Apus were often venerated as the most important of deities throughout the empire (Reinhard, 1985; Gullberg, 2020).

Similar to sacred mountains, many rock outcrops were also thought to be hierophanies, or manifestations of the sacred. Pachakuteq believed that he could improve these stones and, as the son of the Sun and co-creator of the land, he could enhance the work of the creator (Gullberg, 2020).

Camay

The Inkas believed that all things had a point of vitalization, or *camac*. Camay was a concept of specific essence and force, *'to charge with being, to infuse with species power'* and a camac was one "who charges the world with being" (Salomon & Urioste, 1991). The camac for llamas was a dark constellation in the shape of a llama that was responsible for giving the vitality that allowed terrestrial llamas to thrive (Gullberg, 2020; see Fig. 9.5).

Running water was an energizing and animating life force in Andean cultures and so it was associated as an agent of *camay*. In the cosmology described in the *Huarochirí Manuscript* (Salomon & Urioste, 1991) life is born from the feminine Earth by the embrace of masculine water, like how plants grow from soil when moistened by water. The circulation of running water and the pouring of offertory liquids was thought to animate certain inanimate objects which then became waqas and sentient beings with extraordinary powers. Running water was located near many waqas and this further supports that camay was thought to energize the life forces within (Gullberg, 2020).

The world's water cycled through the heavens and Earth as it flowed down the Willqamayu (Vilcanota River) and returned through the Milky Way (Urton, 1981). Inka cosmology described the Milky Way as being a river that flowed across the night sky in a very literal sense. Earthly waters were drawn into it in the heavens and then later returned to Earth after being celestially rejuvenated. The Earth floated in a cosmic ocean (Urton, 1981). When the

Milky Way's orientation was aligned so that it dipped into that ocean, the Earth's waters were drawn into the sky. "The Milky Way is therefore an integral part of the continuing recycling of water throughout the Quechua universe" (Urton, 1981, p. 60; Gullberg, 2020).

Salomon and Urioste (1991) describe waqas as living, energized beings brought to sentience by the Earth's waters. The powers of camay were great, as objects, stones, and even places could be animated with running water or offertory liquids. Snow capped Apus were especially revered because the Inkas recognized them as the source of water in the cycle (Salomon and Urioste 1991). The Inkas also believed that rock could be empowered and energized through elaborate carving (Paternosto, 1996). Carved waqas were given life through the circulation of the Earth's life-force, its running waters (Gullberg, 2020).

Sacred Animals

Many animals, both living and symbolic, were revered and worshipped in Andean culture. The Inkas venerated the Quntur (condor), the puma, and the serpent because they represented the three cosmological worlds of the sky, Earth, and underworld (Gullberg, 2020).

Urton (1981) describes Polo de Ondegardo to have said that "in general, [the Inkas] believed that all the animals and birds on the earth had their likeness in the sky in whose responsibility was their procreation and augmentation" (p. 169). The Inkas' dark constellations in the Milky Way include a serpent, a toad, a tinamou (a bird), a llama, a baby llama, a fox, and a shepherd, the creatures exerting supernatural influence over their terrestrial counterparts (Urton, 1981; Gullberg, 2020; see Fig. 9.4).

Ancestors

Julio Tello, a Peruvian archaeologist, described ancestor veneration as being a significant feature of Andean civilizations (DeLeonardis and Lau 2004). Certain waqas were sometimes shrines honoring ancestors who could influence the living. Feeding of these waqas was a prime motivation for communication with the ancestor-gods (Benson and Cook 2001). Mummies of Inka emperors were also deified ancestors (Zuidema, 1983). They would be carried in processions or displayed on platforms, carved steps, niches, and altars. Shamanic communication with the supernatural world of ancestors and movement between the three cosmological worlds were interlinked (Eliade, 1972). Waqas were places where ancestors could be called upon for assistance with such as agriculture, warfare, health, and fertility (Gullberg, 2020).

Mummies of ruling Inkas played a significant role in Inka culture and worship of them was a part of the state religion. These preserved royal bodies were treated as if still they were still alive (Niles, 1987). When a ruling Inka died his panaqa continued to manage his property and wealth as if he were still alive. His son, the new Inka, had to build his own palace and establish his own wealth. The mummified emperor continued to occupy his palace; he was clothed and 'fed', and he was consulted on matters of significance. The Inka's descendants remained responsible for the mummy and his possessions and would help the mummy participate in state ceremonies (Hemming and Ranney 1982; see Fig. 3.2). Mummies sometimes were called upon to visit other mummies, or even the living (Niles, 1999). All mummies of the ruling Inkas were paraded at a new coronation publically displaying the dynastic lineage of the empire (Gullberg, 2020).

Royal Marriage

A ruling Inka had many wives, but only one was primary. It was customary that he marry his own sister because then he and she would belong to a lineage that was both patrilineal and matrilineal and therefore the emperor would preside over a social hierarchy similar to that of the creator Wiraqocha (Zuidema & Quispe, 1973). They married on the day of the Inka's accession and their children also became primary, the eldest son normally heir to the empire (Zuidema, 1964). Primary children were born by the primary wife and subsidiary children by subsidiary wives (Gullberg, 2020).

Moieties, Suyus, Panaqas, and Ayllus

The Inkas divided Qosqo, and other locations, into upper and lower halves, or moieties. These divisions were both geographical and social. The upslope half of Qosqo was *hanan* (upper), while the lower half was *hurin* (lower). Hanan Qosqo held higher status than did Hurin Qosqo (Zuidema, 1983). There were also social parallels with five royal descent groups assigned to Hurin Qosqo and five more to Hanan Qosqo (Zuidema, 1964). Gasparini and Margolies (1980) suggested that upper and lower divisions may have originated in the mountains due to the need for coordination between ecological zones. Qosqo and the Tawantinsuyo were divided into quarters, or *suyus*. These began in Qosqo and were called *Chinchaysuyu, Antisuyu, Collasuyu,* and *Cuntisuyu* (see Fig. 2.1). The seq'es of Qosqo were organized within these suyus (Gullberg, 2020).

A Panaqa was the patrilineal royal descent group of a newly installed Inka. The Panaqa supported the emperor and was typically led by his second son because the first son would eventually leave the Panaqa when he succeeded his father. As the successor to the throne, he inherited the empire, but not his father's wealth or possessions. A Panaqa took on new responsibility when the Inka died as it became responsible for the care and ceremonial functions of his mummy (Niles, 1999; Gullberg, 2020).

An Ayllu was a non-royal extended kinship group from the same patrilineal ancestor and it provided structure for marriage and inheritance. It served as the basic social unit after that of immediate relatives (Niles, 1987). There were ten Ayllus, five in Hanan Qosqo and five in Hurin Qosqo, and they were paired with seq'es as were the royal Panaqas (Sarmiento de Gamboa, 2009 [1572]; Zuidema, 1964). Panaqas and ayllus responsible for the care of their associated waqas. Seq'es were normally arranged together in threes in groups called *Collana, Payan,* or *Cayao* (Zuidema, 1964; Gullberg, 2020).

Money

The Inkas lived in a society without personal property or money. The state provided for them, and it decided what labor they would contribute for the common good. After the Spanish conquest it was difficult for the Inkas to grasp the concept of earning money to spend for their needs (Hemming, 1970; Gullberg, 2020).

Records

The Inkas never developed a system of writing and they therefore have no traditionally recorded history. All dates and events prior to the Spanish conquest have been gathered through oral questioning. Spanish chroniclers attempted to capture culture and history through interviews with Inka citizens, but often did not really succeed because they were unable to truly comprehend Inka culture (Lee, 2000; Niles, 1987). The Inkas considered it very important to preserve royal history, thus oral traditions were recited during festivals that told of dynastic greatness. Non-royal family history was similarly passed from one generation to the next. Niles (1999) said that "…the deeds of an ancestor were related to the prestige accorded his living descendants…" (p. 24). These oral traditions exist to this day and are used in chapters of this book (Gullberg, 2020).

Khypus

The system developed by the Inkas for somewhat more permanent recording was that of the Khypu, a memory aid with colored strings knotted in a certain order. They developed this method of record keeping instead of a written language to maintain reccords of production, storage, distribution, census data, and taxes (Niles, 1987; Paternosto, 1989). Gary Urton describes Khypus to be a limited substitute for writing, perhaps a hybrid between that and a mnemonic aid (Urton, 2003, 2011; Gullberg, 2020).

A limitation of the Khypu was the likelihood that only its maker could fully interpret the information that had been recorded. The Inkas had a special class of civil servants, *Khypucamayocs*, who were trained as experts in recording and interpreting Khypu data (see Fig. 3.3). The Khypucamayocs therefore had great influence over the content and meaning of Inka official records. Khypus were also used to record calendrical information and as well could serve as a mnemonic aid for family oral histories (Niles, 1987; Paternosto, 1989; Zuidema, 1977; Sherbondy, 1992; Gullberg, 2020).

Textiles

Another form of preserving elements of a culture's history is with their textiles. Weaving preceded even ceramics as an art form in the Andes. Textiles dating as far back as 3000 BCE have been recovered that display the condor, puma, and serpent—the sacred creatures venerated four millennia later by the Inkas (Gullberg, 2020).

Festivals

Pachakuteq's calendar used the position of the Sun on the horizon for times of planting and harvesting, and for religious celebrations. Prominent festivals were Intirraymi at the June solstice, Qhapaq Raymi Quilla at the December solstice, Pacha Pucuy Quilla at the March equinox, and Qoya Raymi Quilla

Fig. 3.3 A Khypu and Khypucamayoc for the calendar. (Reprinted with permission from (Guaman Poma, 2010). © 2010, University Texas Press. All rights reserved)

at the September equinox. Intirraymi (literally Sun festival) was the Inka's winter solar festival at the time of the June solstice and it is still celebrated with great ceremony in Qosqo today. de la Vega (1961) [1609]) wrote that Intirraymi was most important. It was a festival for the masses, and it brought many pilgrims to Qosqo (Dearborn et al., 1987). An elaborate ritual took

Fig. 3.4 Author Milton participating as Kawsaypacha Kamayoq, The Inka Astronomer, in the Intirraymi ceremony; this photo was taken in the Qorikancha before the event began. In his right hand he holds a sundial and in his left a yupana

place over many days; thanks were given to the Sun and prayers said for the crops. Inka citizens chanted throughout day, and they raised the volume of their voices as the Sun rose higher and decreased it again when the Sun later lowered in the sky (Zuidema, 1986; Gullberg, 2020; see Fig. 3.4).

Qhapaq Raymi was a summer festival of the Sun at the time of the December solstice. This was a celebration for crop germination at the beginning of the season. It also was the annual time for Inka adolescent males to undergo the rituals of coming-of-age (Hemming, 1970). Qhapaq Raymi was primarily celebrated by the nobility rather than the masses (Dearborn et al., 1987; Gullberg, 2020).

Climate

The rainy season in Qosqo is from October to April, followed by the dry season from May to September. The Inkas planted maize at the beginning of the rainy season and harvested when the dry season approached (Urton, 1981). Weather patterns were occasionally disturbed during El Niño years. Andeans learned to predict this by assessing the relative brilliance of the heliacal rise of the Pleiades in early June. A dull appearance would predict of a good year for crops while a bright appearance predicted drought (Orlove et al., 2000; Gullberg, 2020).

The Inkas cultivated produce in optimal environmental zones. They learned that certain crops thrived at high altitudes, while others needed a lower climate. Agricultural calendars denoted the different times of proper planting and harvest for each of the zones (Urton, 1981; Gullberg, 2020).

Agriculture

Agricultural terraces worked well for the Inkas, and they used them extensively (Fig. 3.5).

The Inkas adopted the concept from the Huari, who learned this from even earlier cultures (Wright & Valencia, 2000). Terraces were intended to maximize crop productivity and efficiency in mountainous environments; they also protected against erosion and assisted in irrigation. Observations of the Sun and Moon were used to determine the dates for sowing. Maize was the most important crop and pillars were built on Qosqo's horizon to establish the proper time for its planting (Bauer and Dearborn 1995; Urton, 1981). The growing season for maize is from October to May, with the first fields plowed in August (Gullberg, 2020).

3 Facets of Inka Culture

Fig. 3.5 Terraces below Pisac

Irrigation

The Inkas were masters of hydraulic engineering and were adept at irrigating their agricultural terraces and fields. They designed fountains to be both practical and ceremonial, as well as to facilitate the life-force energizing effects of camay. A series of 16 fountains at Machu Piqchu were practical for utility, while fountains at Ollantaytambo and Tipon appear to be more ceremonial. Canals were fed by streams, springs, and reservoirs. Many were lined with stone or were carved into walls. Terraces were irrigated with channel systems and sometimes were backed up with secondary channels (Niles, 1987; Gullberg, 2020).

Pilgrimage

Pigrimage took place in the central Andes long before the Inkas. Early examples in the time of the Huari and the Tiahuanaco can be found at the temple of Pachacamac on the coast near Lima and on the Isla del Sol in Lake Titicaca.

These sites figured prominently in Andean mythology and cosmology and became central points for ritual movements (Zuidema, 2008a). As Inka pilgrims neared the pilgrimage shrine what they experienced had been carefully orchestrated by state and religious officials. The solar cult was featured with its association of the Inka as the son of the Sun. The experience allowed persons making the pilgrimage to transform from residents of mountain villages and become members of the greater empire (Bauer and Stanish 2001; Gullberg, 2020).

Zuidema (2008a, b) discussed three forms of pilgrimage that were practiced along the southeast to northwest axis of the solstices that extended between Vilcanota and Ollantaytambo. He called them (1) a procession, (2) a pilgrimage, and (3) a race. The first two were conducted by Tarpuntay priests near the time of the June solstice while the third involved younger men, perhaps of common status (Zuidema, 2008a, b). In the *procession* priests traveled from Huanacauri on the southeast side of Qosqo to Quiancalla on the northwest. They sacrificed lambs at dawn, at sunset, and along the way at the Qorikancha at noon (Zuidema, 2008a, b; Gullberg, 2020).

In the *pilgrimage* another group of Tarpuntay priests traveled southeast from Huanacauri and Mutu to Vilcanota and then returned. Their journey lasted a month, and they worshipped the Sun while moving through the mountains to Vilcanota, with return by the Willqamayu (Vilcanota River) to Quispicancha (Tipon). This ritual took place in part to arrest the descending path of the Sun across the sky as the June solstice approached (Zuidema, 2008a, b; Gullberg, 2020).

Zuidema said that a foot race called *Mayucati* complemented the pilgrimage. In the race young men competed after the December solstice by running alongside sacrificial ashes cast upon the Willqamayu as they flowed northwest to Ollantaytambo. They then raced each other through the mountains back to Quiancalla and Qosqo in an event set to end at the time of the zenith Sun in February (Zuidema, 2008a, b; Gullberg, 2020; see Fig. 3.6).

Architecture

Each Inka had a palace in Qosqo maintained by his Panaqa, even long after his death. Pachakuteq, Topa Inka, and Wayna Qhapaq constructed fine palaces in the capital and other estates in the countryside. Royal architecture was used by Inka emperors to display status as rulers and demonstrate inclusion in the dynastic succession since the time of Manqo Qhapaq. Rules of succession left a new emperor none of his father's possessions and required him to build

3 Facets of Inka Culture   29

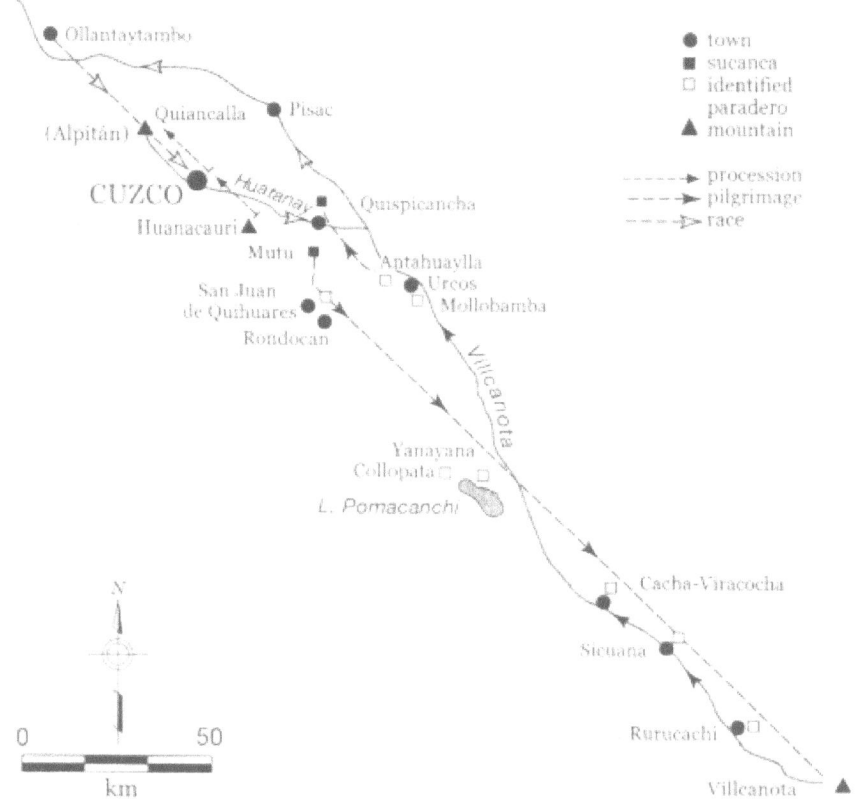

Fig. 3.6 Routes of travel for the procession, pilgrimage, and race. (Reprinted with permission from (Zuidema, 2008b). © 2008 Springer Science+Business Media, LLC. All rights reserved)

his own palaces and country estates and amass his own wealth, a system that allowed for great expression by each new Inka (Gullberg, 2020).

Most civil works of the Inkas were built between 1450 and 1527. Pachakuteq established the empire's architectural style, and it was used throughout the Tawantinsuyo (Gasparini and Margolies 1980; Christie, 2007). Topa Inka used his father's basic model, but with innovations of his own. His son, Wayna Qhapaq, did likewise.

The degree of craftsmanship in the masonry used in a building directly indicated the structure's relative importance. The finer the cut and fit of the stones, the higher the stature of the building. Double-jambed doors denoted entries for use by elite (Hemming and Ranney 1982; Gullberg, 2020; see Figs. 3.7 and 3.8).

Fig. 3.7 Main entrance to Machu Piqchu. (Watercolor printed with permission. © 2024, Jessica Gullberg. All rights reserved)

Inka masonry was generally of two types, polygonal and coursed, and both fit very precisely without mortar (Paternosto, 1989). In polygonal masonry random interlocking faces of stones were cut and polished to fit together with

Fig. 3.8 A double-jambed doorway at Machu Piqchu

precision, and no two walls were the same. Coursed masonry consisted of polished surfaces laid in precisely fitting horizontal rows, each successive row slightly smaller than the one below (Gasparini and Margolies 1980; Hemming and Ranney 1982; Gullberg, 2020; see Fig. 3.9).

Fig. 3.9 An example at Machu Piqchu of a wall carved to fit with great precision and no mortar

Inka structures were almost always single storey with roofs that were thatched. When a second storey did exist, the stairway was generally outside. Common residences were primarily single-room and if multiple rooms did exist, they generally were not connected internally. Doors typically were located on a long wall (Gasparini and Margolies 1980). Qollqa (storehouses) were often circular, as were funerary structures called Chullpas. The Sunturhausi in Qosqo was circular with windows, a high roof, and a mast. Figure 3.10 is a drawing of the Suntuhuasi by Guaman Poma (Gullberg, 2020).

Niches were common on the inside of building walls (see Fig. 3.11). Niles (1987) described them as being symmetrical on walls that faced doorways and many were of a standard size, however some were much larger to accommodate mummies. Stone lintels were placed on the tops of niches and those found on exterior walls sometimes included double or triple jams as signs of status. Niches were often used to display items for religious veneration.

Fig. 3.10 Drawing depicting the Sunturhuasi. (Reprinted with permission from (Guaman Poma, 2010). © 2010, University Texas Press. All rights reserved)

Fig. 3.11 Niches in a wall at Tipon

Qhapaq Ñan

The Inkas built an impressive system of Qhapaq Ñan (roads) for movement throughout their empire that were constructed of carefully fitted stones. The network included more than 16,000 km of rock pavement and many Qhapaq Ñan were built high on the sides of mountains to facilitate travel during the rainy season (see Figs. 3.12, 3.13, and 3.14).

Qhapaq Ñan were built for travel by foot since the Inkas did not have horses or wheeled vehicles. Llamas could be used as pack animals but were not suitable for riding. Carts would have been impractical on such mountain highways, so the Inkas never invented the wheel. Llamas, however, were well suited for transporting goods on their backs along the Qhapaq Ñan. Inka bridges were mainly made with rope (Hemming, 1970; Gullberg, 2020).

3 Facets of Inka Culture   35

Fig. 3.12 Qhapaq Ñan were built throughout the Tawantisuyo

Fig. 3.13 A section of Qhapaq Ñan near Machu Piqchu

Fig. 3.14 The Qhapaq Ñan here is the green horizontal line (vegetation) across the cliff face; this leads to the left to an alternate entrance to Machu Piqchu

Carved Rocks

Examples of carved rock can be found throughout much of the Inka empire. Emperors felt it their right to improve upon nature by sculpting outcrops that often became waqas. The Inkas learned that they could shape blocks by using harder stones (Gasparini and Margolies 1980; Paternosto, 1989; Van de Guchte, 1990; Gullberg, 2020; Fig. 3.15).

Carved rocks were also used to promote ideology and the solar religion. A rock, once carved, became a hierophany and was worshipped by the Inkas. Embedded in the Earth, these sculpted rocks were thought to be connected with the powers of the Ukhupacha (underworld) and became venerated after they were carved. Sculpted outcroppings, as waqas, were important regarding the three worlds of the Inkas (Paternosto, 1989). Ritual stairs are common, expressing movement from Ukhupacha to Kaypacha (the Earth) to Hananpacha (the heavens). These carved stairs normally included three steps, corresponding with the three worlds, and were symbolic representations of this cosmology. Carvings of Quntur (condors), pumas, and serpents were also common as representatives of these worlds (Gullberg, 2020).

Fig. 3.15 Intricate carvings in the Principal Stone at Saywite. (Watercolor printed with permission. © 2024, Jessica Gullberg. All rights reserved)

Summary

To gain a true appreciation of Inka astronomy it is first necessary to gain a significant understanding of their culture. We must not make the same mistake as did the sixteenth century Spaniards by interpreting what we see through our own frame of reference.

The Inkas had a special relationship with the Sun. It formed the basis of their religion, and they believed their ruler to be the son of the Sun. The Moon was also looked upon with great reverence and was thought to be both the wife and sister of the Sun and, by extension, the mother of the Qoya.

Solar worship reinforced the divine authority of the emperor. Shrines and pilgrimages to them were created to ensure the population understood without question the royal relationship between the Sun and the ruling elite.

Mountains and rock were worshipped for their spiritual powers. Deities within snow peaks held the power of life-giving water and supplied it in the cosmological cycle to the people below. Camay was thought to bring sentience to otherwise inanimate objects. Outcroppings were carved to further energize them while enhancing the work of the creator.

The Inkas' cosmos included three worlds, Ukhupacha (the below), Kaypacha (the here and now), and Hananpacha.

(the above), and these were represented respectively by the serpent, the puma, and the Quntur (condor). Carved stairs with three steps were common motifs symbolizing the transition between the three worlds.

Pillars were constructed to use the Sun's horizon position to determine the proper times to sow and harvest crops. Festivals of the Sun were established to commemorate these dates, which also served to reinforce the belief that the Inka had a special relationship with the solar deity.

Many carved stone waqas survive to this day. The Inkas believed that carving and camay energized a spiritual force giving waqas sentience. The Inkas did not view astronomy as a separate entity, but instead as an integral part of their culture. It is only through an understanding of such interrelationships that we can begin to fully comprehend what the Sun, Moon, planets, and stars meant to the peoples of the Andes.

4

Inka Culture and Quechua Language

Contents

Time, Space, and Language	42
Kawasaypacha	43
Ayni	44
Tawantinsuyo	45
Inka	48
Ayllu	49
Panaqa	49
Waqa	49
Apu	50
Apus and Kawsaypacha	50
Apus, Waqas, and Ayni	50
Ushnu	51
Minka and Mit'a	51
Sumaq and Allin	52
Tinkuy	53
Iskaynintin	54
Masintin and Yanantin	54
Mass	55
Yana	55
Llaqta	56
Paqarina	56
Inka Creation	56
Usufruct	57
Currency	58
Summary	58

The original version of the chapter has been revised. A correction to this chapter can be found at https://doi.org/10.1007/978-3-031-67580-5_12

© The Author(s), under exclusive license to Springer Nature Switzerland AG 2024, corrected publication 2025
S. Gullberg, M. Rojas Gamarra, *Inca Cosmovision*, Astronomers' Universe, https://doi.org/10.1007/978-3-031-67580-5_4

To understand Inka astronomy and cosmovision it is important to have an understanding of Inka culture. Many languages existed in the Tawantinsuyo and still do today. The main one was *Quechua*, or *Runa Simi*, and a familiarity with terms in the Quechua language is necessary to begin to better conceptualize the worldview that existed throughout the Tawantinsuyo. This chapter will introduce certain Quechua terms, and doing so will help greatly to embrace the logic of life in space and time in Inka culture. Main principles of life such as Ayni, Kawsaypacha, Iskaynintin, Mit'a, Minka, Masintin, Yanantin, and Tinkuy will be explored (Fig. 4.1).

The Inkas benefitted from knowledge learned from the many cultures that came before them. In their expansion they did not destroy the cultures they annexed; to the contrary, they improved their "empire" by assimilating knowledge and incorporating it into daily use. In order to better understand Inka culture, its *cosmovision* (the logic its people used to think about the world in the context in which they lived), its way of life, and how it developed, you must be immersed into the world of the Inkas.

Fully understanding a culture using only words is not really possible. It is difficult to accurately explain certain feelings, smells, flavors, etc., and elements of Inka culture are more difficult to explain in languages other than Quechua. Inka culture is introduced with Quechua termas and good approximations in English, but since words in English still can bring Western connotations, as much as possible Quechua terms will continue to be used once they have been introduced. This chapter comes from the Quechua language and culture that author Milton has experienced thoughout his life, and through the Inka oral traditions passed on to him by his elders.

All Tawantinsuyan knowledge was interrelated, be it regarding economy, justice, politics, education, routines, or astronomy. Understanding the logic of life by knowing certain terms in Quechua will help add perspective for a better sense of Inka astronomy.

In the Inka world everything was integrated and related, this due to a principle of life called *Kawsaypacha*, a concept best summarized by saying that "everything in the cosmos lives." This ethos affected all aspects of the worldview of the Inkas, including astronomy, and is why it is important to grasp because doing so will help to better understand the role that astronomy played in their culture. This also helps to illustrate how the Inkas viewed their origins and the cosmos.

Everything in the cosmos lives—the Sun, the Moon, stars, constellations, planets, hills, Apus, Paqarinas, and Waqas all were sentient.

4 Inka Culture and Quechua Language

Fig. 4.1 Author Milton as Kawsaypacha Kamayoq, the Inka Astronomer, at Intirrami

Time, Space, and Language

Time in the worldview of the Inkas was not separate from space, it was part of the Kawsaypacha principle in which everything in the cosmos lives. Once again, in Quechua, the word Kawsaypacha means that space-time lives. Modern Quechua dictionaries can be problematic, though, in that they lose oral tradition in what has been written. This is because the context of many words has changed over time and that pronunciations have varied as well. There also are different dialects of Quechua in Ecuador, Peru, Bolivia, Chile, and Argentina. Yet another factor is that Inka Quechua was not a written language. Following the Spanish conquest writing it was attempted by Spanish chroniclers who employed similar sounds in Spanish, and these ambiguities have contributed to the multiple spellings of Quechua words seen today.

Amauta is a Quechua word meaning a wise and noble man, a teacher and keeper of knowledge. Amauta Emilio Huaman Huillca gives example with the letters K and Q that he says lead to confusion as how to spell "huaca." In Quechua words exist such as "wak'a" (crazy), "waqa" (sacred place), or "waca" (cow). His answer is waqa and there are many other spoken words that experience similar confusion of spelling when they are written.

Another consideration is that written Quechua words have been created using definitions of concepts or metaphors. An example of this is the word "phuyupatamarka" where "phuyu" means cloud, "pata" is the high part of something or the top of a hill, and "marka" is a city, town, attic, or protector. This might lead to a translation such as "town of the clouds on the top of a hill." Amauta Emilio says that there are even Quechua words not known in the time of the Inkas that have been created more recently.

Several contexts exist for the word "pacha." On one hand pacha means world or Earth, such as with Kaypacha—this world, Ukhupacha—the world within, and Hananpacha—the world above, the three cosmological worlds of the Inkas. However, pacha can also be used as a suffix that indicates the moment, the instant and time of action. Examples of this are kunan pacha—at this moment, and haqaymanta pacha—from that place. With kunan pacha the word pacha means time and with haqaymanta pacha it means place. Both come together in the concept of space and time.

When pacha is used for the world, the Earth, or the cosmos it is more concrete. The Spanish chronicler, Pedro Cieza de León (1998 [1555]) described pacha as it related to "knowing the turn that the Sun makes and the waxing and waning of the Moon." Pachamama, Mother Earth, can be described as serving to fulfill Kawasaypacha. Pachayachachipa includes "yachay" which means knowledge and wisdom, and pachacamacpa includes a variation of "kamaq" meaning creator or inventor. This shows the word pacha as being associated with "creator" and "knowledge," supporting pacha both as world and as space-time. When pacha is used as a suffix it almost always refers to space-time and when used as a single word it means world, Earth, or cosmos.

In short, the word pacha can be as abstract as space-time or as concrete as the Earth itself or the cosmos. Time and place are presently thought of separately, but to the Quechua the two are the same with pacha. This is a different perspective from that of the Western concept of time and space.

The Inkas saw time as a place. Everything was integrated and related and the word pacha is one of the examples that support this concept.

Kawasaypacha

Again, Kawasaypacha can be described by simply saying that everything in the cosmos lives. The Sun, the Moon, the stars, the constellations, the planets, the hills, the apus, the paqarinas, and the waqas all live (see Fig. 4.2).

The word Kawasaypacha combines two Quechua words, *Kawsay*, which means living, and *Pacha,* which means space-time at a time, and together this means "space-time lives." There are many words in Quechua that contain the

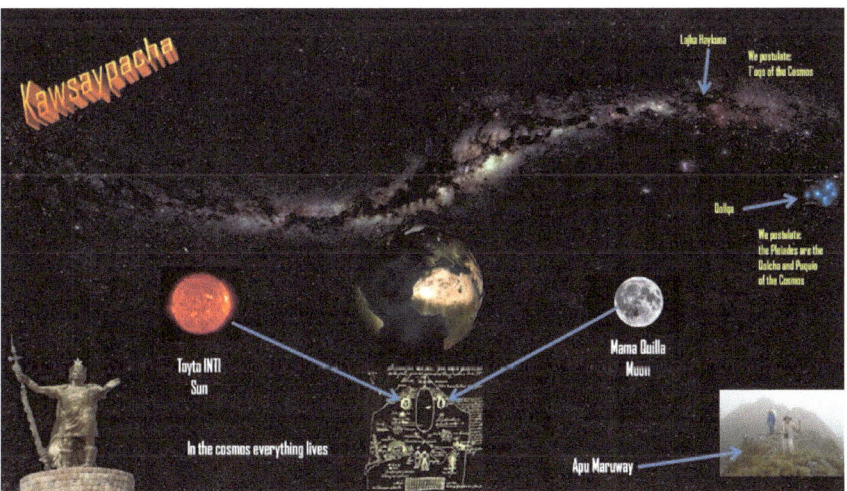

Fig. 4.2 Kawsaypacha—everything in the cosmos lives. Included in the upper right are the constellations Lajha Haykuna and Qollqa, otherwise known as the Pleiades. The Inkas thought that all of the cosmos came from these two places. (Modified with permission from (Magli, 2004). © 2005, Birkhaeuser Verlag AG. All rights reserved)

word Pacha, such as Pachah, Pachahchaky, Pachakamaq, Pachakuteq, Pachamita, Pachamama Pachatusan, Pachan, Pachak, and Pachaq. Another reason that time and space have been joined in the word Pacha is because Andean people have always measured time using space. Solar pillars were placed on top of hills to mark when the Sun entered times of special solar horizon events. Thus, time was measured using these pillars called *Sukanqas*.

Ayni

Ayni was the main principle of life throughout the Tawantinsuyo, for all including the nobility. Ayni is commonly translated as "reciprocity." The essential characteristics of this reciprocity are as follows:

(a) Ayni was fulfilled.
(b) Ayni was in the collective imagination; it was the way of being and living in all activities such as Mit'a and Minka.
(c) Ayni was not fulfilled by obligation, or by commitment; consequently, the Ayni was simply fulfilled.
(d) Ayni was kept without awareness.
(e) Ayni was also fulfilled with all of the Apus and on Earth and those in the firmament. This is because Kawsaypacha was fulfilled. This fulfillment was carried out in large meetings called Tinkuy, which were also performed as a principle of life.

It can be said that Ayni was a natural law. In the Tawantinsuyo, the territories and cultures annexed to the Inka state normally were not forcibly subjected. Only with peoples who fought the Inkas, such as the Chankas, were there conflicts, but in the end even they cooperated. In the Tawantinsuyo there was no imposition, there were only agreements that were made with the annexed Llaqtas (towns, areas), and thanks to the principles of life, the activities carried out by the inhabitants were of their own free will, without force, all in Ayni.

Even the *Tinkuy* (meetings) were a mixture of ceremony, activity, and joy in which all sang and danced. Modern "work" did not exist; contemporary conceptions come from the word "work," and it must be understood that in the Tawantinsuyo activity was done voluntarily, with vocation, in ceremony, and in complete joy. In the Tawantinsuyo there was no hegemony; it could not exist because of Ayni.

Inclusion or exclusion using possessive pronouns refers to the one who is speaking. For example, the chief of an Ayllu speaks to another chief and tells him "We will work this land." But that chief will not be able to do so because he had to travel to Qosqo. His response would use the exclusive we (us), Noqayku—"Chakrata noqayku llanqayku." But if he would be able to work (inclusive) he would then say "Chakrata noqanchis llanqanchis." With possessives the Inka could say "our lands" without including himself by using the phrase "Kay Pachamamanku," but with him included it would be "Kay Pachamamanchis." From this example, the Inka could speak in an exclusive way about a land annexed to the Tawantinsuyo. He had only to enforce the principles of life, which was not difficult for him to do because everyone had them internalized. They lived by these principles daily.

Living in concert with such principles of life is thought by some to have first developed with the early humans that migrated to South America. It is thought that the last cultures to form were those in South America because this was the last place that human beings reached after leaving Africa, crossing Asia, continuing to North America and then on to South America. These people needed a great bond to endure such a tremendous journey, and this was ingrained within them. Such has been called the explorer gene and it is said to have prompted humans to take these risks; succeeding thanks to their unity. This long process of social coexistence and evolution of beliefs may have led to Inka life principles such as Ayni, Mit'a, Minka, Tinkuy, and Kawsaypacha.

As an example, it must be noted that newly joined couples were assigned land, a topo to the woman and a topo and a half to the man. *Topo* is a measure of land area that is thought to be between 0.2 to 0.33 hectares, according to oral sources. They would build their houses there, and they did it in Ayni with the help of the community interaction of Minka. In Minka, land was worked by all, and the products were for the whole Ayllu. Thus, there was no need for private property. If this practice of reciprocity did not exist it would cost a couple much more to build their house and till their land.

Tawantinsuyo

Language can be thought of as a product of the customs and beliefs of a society corresponding with the environment in which it exists. A language is born out of necessity when interacting with the world and the surrounding environment. The Inka landscape belongs to the mountain range called the

Coordillera de los Andes and this environment significantly influenced the development of the Tawantinsuyo language and culture.

Once again, the Tawantinsuyo is the region where Inka expansion thrived from its beginning in the early 1400s until the Spanish conquest that began in 1532 CE. The Inkas spread across the Coordillera de los Andes range to territories that are now in Colombia, Ecuador, Peru, Bolivia, Chile, and Argentina. Further Inka influence is thought by some to have extended to parts of Paraguay, Uruguay, and Brazil. It is thought that to the east of the of the empire there extends to the Brazilian coast a section of Qhapaq Ñan, the Quechua name for the system of Inka roads.

The Inkas also moved south into the region of the *Mapuche* in Chile and Argentina. They fostered a cultural exchange with them with no military confrontation. Similar festival traditions exist in both cultures, such as the Mapuche *Witripantu* which occurs at the same time as the Inka *Intirraymi* at the solstice in June. There also are Quechua words that exist in the Mapuche lexicon.

It has been speculated that Polynesia may as well have been reached by the Inkas with a trip made by Tupaq Yupanqui, and this possibility is thought by some to be supported by Quechua names of certain islands and a wall on Easter Island that very much resembles an Inka wall.

Tawantinsuyu is the union of two Quechua words, "tawa" meaning "four" and "suyu" meaning "region," and therefore describes the four regions (see Fig. 3.4) with which the Inkas divided their "empire" with the center at their capital of Qosqo (Cusco). Qosqo in Quechua means "navel," not only because it was more or less at the center of the empire, but also because of what a navel represents with mammals. It must pointed out, however, that this was not really an empire. The word "empire" (from Latin *Imperium* meaning dominion imposes) implies two fundamental characteristics:

(a) a state ruled by an emperor (who has power and dominance, the one who imposes and dominates).
(b) a set of states or territories subject to another.

Empire is a political organization in which a state or nation imposes its power on other countries—hegemony supremacy that a state or a people exercise over another. Look at the differences:

(a) In the Tawantisuyo there was no emperor, this term was introduced by Europeans. The Inka as the leader was respected not by force but through the principle of life called Ayni, which included the basis of social interactions called Mit'a and Minka. In Ayni everything in the world is connected and the people share with reciprocity and mutualism. This meant that because of Ayni the Inka could not dominate or impose, and his power was not through force and imposition. Power exercised by the Inka was that of willing obedience.

(b) In the Tawantinsuyo the territories and cultures annexed to the Inka state normally were not forcibly subjected. Only with peoples who fought the Inkas, such as the Chankas, was there conflict. In the Tawantinsuyo there was no imposition, there were only agreements that were made with the annexed Llaqtas and, thanks to the principles of life, the activities carried out by the inhabitants were of their own free will, without imposition and in Ayni. Even the tinkuy (meetings) were a mixture of ceremony, activity, and joy in which all sang and danced. Work was done voluntarily, with vocation, in ceremony, and in complete joy. In the Tawantinsuyo there was no hegemony; there could not be hegemony because Ayni existed and was fulfilled (see Fig. 4.3).

"State" can be said to have a set of powers and as being the governing body of a sovereign country. The difference between a modern state and the Inka state is in how it was governed. In saying "govern" it can be easy to envision powers of the state to govern with a bureaucracy. In the Tawantinsuyo, however, the ruling Inka led without the need for a bureaucracy with state agencies.

Fig. 4.3 This is a typical Andean and Tawantinsuyan landscape. It belongs to the Coordillera de los Andes and is where the Tawantinsuyan people lived. Welcome to the Tawantinsuyan world!

The rules of modern states are frequently geared toward preserving privacy for people and protecting their intellectual and material possessions. This was different in Inka culture because almost everything belonged to everyone—there was no private property, there were no patents, and there was no intellectual property meaning that there was no need for attachment to material things. So how do you govern such a state? In the Quechua language there is no specific word to express that something belongs to someone, instead you add a suffix. For example, "son" in Quechua is translated as "churi" and to say "my son" is "churiy," "your son" is "churiyky," "his son" is "churin," and "our son" is "churinchis." The Inka being the only "son of the Sun" is expressed in Quechua as "Intip Sapan Churin." Churi means son, Churin means his son, Sapan means unique, Inti means the Father Sun and Intip means of the Sun, therefore Intip Sapan Churin means the only son of the Sun.

A suffix can also be added to personal pronouns. Noqa means I and Noqaq means mine, Qan means you and Qanpa means yours, Pay means him and Paypa means his, Noqanchis means us (inclusive) and Noqanchispa means our (inclusive), Noqayku means us (exclusive) and Noqaykuq means our (exclusive), and so on.

Giant monuments such Saqsaywaman might have been constructed in one of two ways, one with forced labor but the other done voluntarily in Mit'a based on Ayni. The Spaniards took advantage of the interaction called Mit'a but did not do so on the basis of Ayni.

Therefore, this discussion regarding the Tawantinsuyo seeks to understand the spatial place where Inka culture developed and understand the Inka "empire" or the Inka "state," with characteristics such as Ayni as the principle of life. When the word "empire" is used in this book, it is for convenience even though it does not directly relate to the world of the Inkas.

Inka

The *Inka* was the one who led the Tawantinsuyo in Ayni. The relationship of the Inka to his people was very different from that of a king, emperor, or governor. The Inka, like the Tawantinsuyanos (the runes, the people of the Tawantinsuyo), had the principle of the Ayni immersed within him. With this principle governing his life he could not be a tyrant or a despot, this was in principle not possible. The Ayni was ingrained in the Inka as natural law.

Here "Ayni" will be thought of as "reciprocity," and as such it can be said that the Inka practiced reciprocity with the people of the Tawantinsuyo, with all the Llaqtas (towns), and with all the Ayllus.

Ayllu

An *Ayllu* is a set of families united by various ties. These ties could be religious since they had the same Waqa, the same Paqarina, the same Illa, the same language, almost the same job, or the same activity. The Ayllu was the base of Inka social organization and was made up of several "families." A main characteristic of an Ayllu was that it lived using the life principle of Ayni.

Panaqa

A *Panaqa* is an Ayllu of an Inka. It is formed by the Inka, his family, and the people closest to him and all were from the Sana Inka group (Inka nobility). In some chronicles this appears with the name of "Royal Panaqa," but remember that royalty is a Western concept that has nothing to do with the Inka's Panaqa.

The most important characteristic of a Panaqa is that its members also lived in Ayni. The head of the Panaqa was the Inka.

Waqa

Waqa is something out of the ordinary, an amazing entity. It can be a special place, object, animal, or person and it can also be a special physical phenomenon. This can be in the Pachamama (the Earth) or in the firmament (the heavens). Even be Inka constructions could be Waqas.

de la Vega (1961)[1609]) said that a Waqa was "… an admirable thing, worthy of admiration for being beautiful, as it also means an abominable thing for being ugly …. ".

A Waqa could be a hill, a hole, a fountain, a lagoon, a rocky outcrop, a tree, twin children, children who were born different, a protective amulet (Illa-Totem), constellations, the Sun, the stars, the bright planets, the Moon, and the Paqarinas.

Garcilaso also related "… various meanings that this name huaca has, which, pronounces the last syllable at the top of the palate, which means idol, like Jupiter, Mars, Venus, …. ".

The Catholics placed crosses in numerous Waqas and on many of these sites chapels and cathedrals were built. This was done in the Church's extirpation effort as part of converting the people of the Tawantinsuyo to Catholicism. In this campaign some were tortured and some were killed and the people were both manipulated and frightened. They did not leave their Waqas, however, and maintained the relationships with them that had been formed over hundreds of years.

Apu

An *Apu* is a natural guardian, a living being. An Apu is a special Waqa (an Apu is a Waqa but not all Waqas are Apus). It could be said that an Apu was thought of as a god or deity, but there are differences.

In Inka culture an Apu partly corresponds with "God" in the West, but unlike the Western God, Apus can be seen. For example, the Tayta Inti (the Sun) is not supernatural, it is there and you can see it. Apus were spectacular hills and glaciers in the Tawantinsuyo, as well as celestial bodies such as the Apu Taita Inti (the Sun), Apu Mama Quilla (the Moon), and the stars. Apus were asked to protect, guide, support, and defend the people in their activities in the Llaqta, and when they traveled they asked the Apus to keep them well.

Apus and Kawsaypacha

For the Tawantinsuyans, all Apus had life of their own; they were not anthropomorphized, they were not changed in shape, and they were not given human physical characteristics. The Apus had their own physical characteristics, including their own psychological and sociological characteristics. These were their characteristics and their own way of feeling.

Apus, Waqas, and Ayni

Apus and Waqas had life, and so there was a need to communicate with them and to have social interaction with them. This was done in great Tinkuy, the large meetings where very cheerful and pompous ceremonies and parties were held. These interactions were done using all life principles, but mainly the principle of Ayni. The Apus were considered to be important members of the Ayllus, and thus the people of an Ayllu lived in community with the Apus.

With Minka, houses were built for all members of the Ayllu, so because Apus were Ayllu members a house had to be built for each of them.

The Inkas regularly interacted with the Apu Tayta Inti, which was the most important Apu because it gave them life. Apu Tayta Inti gave its energy and warmth. Thanks to the Sun plants grow and thanks to the Sun glaciers thawed for water. The Sun comes out every day and gives light. Thanks to the Sun there are different climates throughout the year. The father Sun needed his house too and that is why Intikancha or Intiwasi was built for him, which is the Qorikancha. In order to talk with Apu Tayta Inti, as well as with other Apus, different Ushnus were erected. And to follow the Sun across the sky Intiwatanas and Sukanqas were constructed; with these the Sun told how tomanage the year for crops. All the people of the Tawantinsuyo interacted with the Sun in Ayni and the principle of Kawasaypacha was fulfilled.

Ushnu

Ushnu is a place or platform where the Tawantinsuyanos (the people of a Llaqta and their leaders) communicated with the Apus and certain Waqas. They were generally platforms oriented to the Apus or to the Ayllu's protective Waqas.

With Kawsaypacha and Ayni, the interaction of the Ayllu with Waqas or Apus took place in large Tinkuy, which again were very pompous meetings with a lot of fraternity and merriment. The Tinkuy had both its ceremonial part and its joyful and fun part (with songs, dances, and jokes). Ushnus were built not only for the Waqas or Apus located in the Kaypacha (in the Pachamama, the surface of the Earth), but also for those located in the Ukhupacha (the world below) and in the Hananpacha (the world above), which is the firmament or the cosmos. Ushnus were located throughout the Tawantinsuyo, and these help now to know how far the Tawantinsuyo extended.

Ushnus needed to be in strategic locations from where the Apu could be observed. The most prominent Ushnus were located at the Qorikancha, where a Cathedral is today in the main square of Qosqo. These constructions reinforced the principle of Mit'a that was a function of the Ayni. This had to be in Mit'a because the Apus were members of the Ayllus (Gamarra et al., 2024).

Minka and Mit'a

Minka is the principle of interaction in Ayni between Ayllu members and a rune (person) or married couples. This premise dealt with "help" or "work" for a newly married couple. A married couple led runes to the house (a tradition that is still practiced with the name of *Minge*) or to land to be sowed or harvested. The Ayllu took turns cultivating the land of all Ayllu members.

Mit'a is the principle of interaction in Ayni between those who directed the empire and the members of the Tawantinsuyo. Ayni is where the leaders of the

State provided a series of benefits such as ensuring housing, food, health, protection, education, knowledge, distraction, and joy, and the people in turn gave community service (Llankay), which served the leaders and the people.

The Ayllu shared in the cultivation of the land of all Ayllu members. The two necessary premises were the Mit'a and the Minka, which were met as natural law.

Activities that were done in community service were such as the construction of Chakas (bridges), the Qhapaq Ñan (great path), Qollqas (store houses), the architectural complexes that served to meet the tinkuy with the Apus, the lands of the Nobility, and the lands of the state. Llankay is closer to the concept of community service than is the word "work."

Sumaq and Allin

There are more principles such as Sumaq Yachay, Sumaq Munay, and Sumaq Kawsay. *Sumaq* and *Allin* are not easy to translate precisely but, more or less, Sumak means splendid, great, in full, or even sweet; Allin means well, but the word "well" should be clarified for its use in the time of the Inkas. It is a good that is quite different from the Western "good" because Western "good" can also imply that there is evil, and that good is rewarded and and evil is punished. Thus, those are categories of domination or serving for domination; doing well Westernly must not be mistaken for that of the Inkas becaue it is not the same as the "good" of the Inkas. For the Inkas if you were wrong, you did "well" too. To do "well" could also be used with wrong and thus a statement such as "for doing well I was wrong" implied that evil did not exist in Inka thought. In the Tawantinsuyo it was only from the heart (Munay), with sincerity, with thought (Yachay), and perhaps with "innocence," but always interacting with Kawsaypacha and among themselves with Ayni, which is in resonance and harmony. Of course, the Inkas could make mistakes, but when they did they were corrected but not punished. If the word "well" is used this must be taken into account. Sumaq Munay, or Allin Munay, is an Inka principle that consisted of wanting splendidly, sincerely, in full, grandiously, sweetly, and to want "well." Sumaq Yachay, or Allin Yachay, is an Inka principle that consisted of thinking splendidly, sweetly, or to think "well" in which Ayni is present.

Sumaq Ruway, or Allin Ruway, is the Inka principle that consisted of doing splendidly, in full or sweetly, or to do "well." The use of Sumaq Munay, or Sumaq Yachay, is wanting splendidly and thinking splendidly. Splendidly, if you want "well" and think "well," then it is done "well." This implies interacting due to Ayni as a natural law, so then the Sumaq Ruway is interacting in Ayni either in Yanantin or Masintin, between Runes (men) and Kawsaypacha, or among themselves. Sumaq Kawsay, or Allin Kausay, is an Inka principle that was fulfilled as a natural law that consisted of living splendidly or in good living. It was based on Ayni and interacting either in Masintin or Yanintin, wanting good (Allin Munay) and thinking good (Allin Yachay), and it is good (Allin Ruway).

Tinkuy

Tinkuy in Quechua means a meeting or interviewing of two or more people. It also means getting to touch two things, it is synonymous with Laythuy and Tupay, and it is also the union of two Andean ecological zones, the Qhehewa and the Puna, at 3500 and 4300 meters above sea level, respectively.

Activities (jobs) were done in Tinkuy, that is, through an interaction with special characteristics, activity instead of work. Tinkuy could have one of the two forms that have been described, Minka and Mit'a, and both functioned with Ayni and Kawsaypacha. Thus, these activities began with a permission ceremony for the Pachamama and the local Waqas and Apus, who were to participate throughout the activities because they were considered to be living entities. All activity was carried out with the characteristics of Tinkuy.

The characteristics of Tinkuy were as follows:

(a) Based upon the principles of life called Ayni and Kawsaypacaha.
(b) The Pachamama, Waqas, Apus, and Paqarinas were acknowledged with ceremonies, because everything was considered to be alive.
(c) All of these living entities "participated" in the activities, for example by having the Pachamama drink chicha before the participants did so.
(d) The activity included singing and dancing with great joy and happiness.

Imagine making jokes with music in the background and a lot of festivity. The author, Milton, has participated in these activities in communities for pro-housing associations where there are tasks to be performed. Jokes emerge in those meetings and there now are discussions as well.

Therefore, the activities (jobs) were performed singing and with great joy. People were not obliged to work the land that was given to them; to the contrary everyone anxiously awaited these activities due to the distraction they would have while carrying them out, and they did this voluntarily. And if they did not work the lands given to them, what would they eat? Because tasks were performed in Ayni (with help) and also with distraction these tasks were not at all tedious. They sang with much rejoicing and happiness.

Using information from the chronicles, current customs, and what author Milton learned from his grandparents, family friends, and from the Amauta Emilio Huaman Huillca, it can be said that Tinkuy was the beginning by which these interactions existed. Regarding the Iskaynintin in space and time which is based upon Ayni, it can be said that Tinkuy existed to comply with Ayni. For example, in a Tinkuy thanks would be given in Ayni (reciprocity) to the Sun since everything in the cosmos is alive (Kawsaypacha). Ushnus were constructed for ceremonies for the Apus. Other examples were Tinkuy made to perform the Minka and Mit'a as rituals. In the case of the Minka, de la Vega (1961) [1609]) said that there were tremendous meetings that were made with a lot of joy and jubilation, in which before, during, and after the realization of the work joy, drink, and food was always present. In the case of the Minka, all the Ayllu or several Ayllus were combined to perform the sowings of crops or to build the houses of Ayllu members. Today this is still carried out to a lesser degree throughout the Tawantinsuyo; even in southern Chile exists the Muff where a town is tasked with moving houses, now with the help of oxen.

Iskaynintin

Another principle in the worldview of the Inkas was the *Iskaynintin* principle of duality. *Iskay* meaning two, double or pair, and *Nintin*, is an inclusive derivational suffix that means with her, according to Amauta Emilio Huaman Huillca. Everything in the cosmos has its dual (Gamarra & Zen Vasconcellos, 2019). In the cosmos, the Hananpacha (outside world) and the Ukhupacha (the inner world) were connected by the present world of Kaypacha for duality.

Duality also has to do with a word pair. If the inclusive derivational suffix Nintin is added, there will be Masintin and Yanantin that are two ways of pairing the dualities (the Iskaynintin) found in the Tawantinsuyo.

Masintin and Yanantin

Masintin is a way to pair the Iskaynintin and these pairings were similar at the time of interaction from equal reference levels. Here the Ayni is the key that makes supplementary union possible; the relationship after the interaction was strengthened and united. Masintin was a generalization (to match) to Minka.

Yanantin was another way to pair the Iskaynintin. These pairings were complementary and not similar according to function. From the interaction between two dualities, a new duality emerged, and this third duality strengthened and united duality. Yanantin was generalization (to match) to Mit'a.

Mass

Mass in Quechua means a couple, there is also the word *Masachaq* which means that it brings together two people. The words Masachay or Maschay mean to match, a pairing, a gathering, or joining two things or two people. Masi can follow the subject and means resemblance, with these words can be formed such as Llaqtamasi meaning countryman, Runamasi meaning similar man, Qharimasi meaning similar, Warmimasi meaning similar woman, and Wasimasi that could mean a similar house but also translates as a neighbor. Masi as an adjective that means equal or similar to.

Masichakuq means that it associates. Masichay means associating or enrolling. Masinakuy means dealing with much confidence or equal treatment of two people and Masani means brother-in-law.

Masintin also is a synonym of Purantin. Pure is a suffix that indicates peers or purely of the same class, species, or form. It is also synonymous with Kama, and there are examples such as Yachaqpura that means between knowledgeable and Llankaqpura meaning among workers. The word Purentin means between the two of the same space, class, category, or form; the words can be combined as Llaqtapurantin Risunchis which together means to go between those of the same people.

Yana

The word *Yana* as a noun, means couple, boyfriend, lover, in love, and also means groom service or servant. As an adjective it means black and is an antonym of Yuraq that translates as white, so Yuraqmasi would be with his white and Yanamosi with his black. Yananchakuy is a verb that means to appear or get married that is a synonym for Saway. Yanantin also means a couple of boyfriends, or a couple of lovers or cohabitants.

Yanapachikuq as an adjective and noun that means help. Yanapachikuy as a verb means getting help or collaborating with another person. Ruranakunapi Yanapachikuy means getting help in chores.

Llaqta

Llaqta is a city, town, village, or community. In a Llaqta, lands were not private property owned by the people or the state. The Inkas believed that the land, the *Pachamama*, was a living entity. She was spoken to, fed, asked for permission to cultivate, monuments were made, and rocky outcrops were carved around stone enclosures. Important parts of the Pachamama were the Waqas (sacred objects), Apus (mountains), and Paqarinas (places of origin and final destinations of the Ayllus). No one could own something that was alive. Today ceremonies are still held to "work" the land, and there are special days of the year, such as the first of August, on which all houses and lands are adorned with yellow confetti and little flowers. There are ceremonial "payments" to the Pachamama in June at Intirraymi and there are other ceremonies each month dedicated to her, and as well as to Tayta Inti (the Sun), Waqas, Apus, and Paqarinas.

Paqarina

Paqary in Quechua means "born," "appeared," "created," or "originated." *Paqarina* is the place of origin where the Ayllu or the Tawantinsuyo Runes believed was the place from where the first of them came—from where the Ayllu originated. Paqarina is a source of life. A Paqarina is a special Waqa (a Paqarina is a Waqa, but not all Waqas are Paqarinas).

The Firmament (the heavens) had Paqarinas, and these were the constellation Qollqa and the T'oqo, which was a dark region in the modern constellation of Argonavis.

Inka Creation

As in all cultures, early Andeans wondered about their origins and also the origin of everything they could see, including the cosmos. According to Inka beliefs, the Ayllus came from Paqarinas. There are two types of Paqarinas, first which have to do with water and its equivalent in the cosmos and second those that have to do with a hole in a geographical or cosmic place. Those with water were called Lagunas Qochas and the holes were called T'oqo. The lagoons chosen as Paqarinas were generally those on the edge of snowfalls.

The Inkas sought their origins and thought that the first of the Ayllu came from the Puquios and Qochas largely because of the importance of water for agriculture. The Inkas did not know about photosynthesis or how nutrients assimilated through roots, they only knew that water germinates a seed and therefore water gives life to a plant. With that reasoning the Inkas thought that the first of its Ayllu should have come from water.

They also thought that the first of the Ayllu was born from a cave or a hole (T'oqo) because of the similarity with functions of holes in the Earth. A plant comes out of a hole, and humans emerge from a woman's vulva. Thus, the first of the Ayllu should also have come out of a T'oqo.

There are two Inka mythologies related to this. First, the legend of Manqo Qapaq and Mama Oqllo, as described by the Spanish chronicles and through oral tradition, where it is said that they came from Lake Titicaca. Second, the legend of the four brothers Hayar emerging from a cavern, a T'oqo, called Cerro Tamput'Oqo.

Similar myths describe the origin of the cosmos. The Inkas had a celestial Puquio (source), which was Qollqa, the Pleiades, because it is blue like water. All water constellations emerged from Qollqa, similarly to water coming from a spring. The Inkas also had a T'oqo (hole) in a dark part of the present-day constellation of Argo Navis, also a Puquio that was called Lajha Haykuna. The two are shown toward the upper right of Fig. 4.2.

Usufruct

The Inkas who directed the different regions of the Tawantinsuyo had platforms made to deliver agriculture to the people within those lands. They were given topos (land moles) when they formed a family, which were temporary and for their *Usufruct* (distribution). According to de la Vega (1961)[1609]), a topo corresponded to a bushel and madia, which was 60 steps long and 40 steps wide—2400 square meters if the steps were one meter, or 1200 square meters if the steps were 0.8 meters—the length of an average step. A man was given a topo of land and his wife a half topo. Other interpretations have said that the man was given a topo when he joined his partner, and the woman was no longer given her half topo because she would now live with her partner. Another plan said that when they had a son, they were given one more topo and when they had a daughter, they received another half topo. It also has been thought that when the male joined his partner to form a family, the male took his topo or was given another topo in another place and the woman lost her topo because she now cultivated and consumed from her partner's topo. It has been said that the lands could be distributed every year, but in general they are thought to have remained constant, because they were well known by those who worked them. The size of the areas of land given show that no one had more land than they required for subsistence; they would work the land for food and could trade for varieties of food that they did not produce.

Currency

Another important characteristic of Inka Llaqtas was that there were no currencies to be used to obtain types of food not personally produced. Instead, people bartered, and all exchanges had their equivalence. Today exchanges are still made in rural areas. As an example, two sacks of potatoes are equivalent to a leg of cattle.

When there is no currency in a village and a person's job is to directly produce one's own food, this engenders a different philosophy of life, in ways perhaps a better one where many problems that societies have today did not exist.

To live well people needed to eat and be happy; to eat you work your land and to be happy you participate in the Tinkuy. This, in part, is what today is called *Sumaq Kawsay*, or "living well, living fully." Sumaq Kawsay is the union of "allin munay" (wants well) and "allin llankay" (works well).

Summary

Inka society was clearly one of farming—most of its activities had to do with agriculture. For good harvests they needed to know two things: first, they needed to understand the climates of regions in different parts of the year, and to address climatic changes the Inkas created calendars related to the cosmos. Second, they had to know what made plants grow.

They thought about the origin of the first seed and the origin of the first of Ayllu, and from that they extrapolated their interpretation of the origin of the cosmos. They thought of water, which is a source of life, as being born in the lagoons on the edge of snowfalls, and they thought that the first of the Ayllu left these places. They also realized that a hole made in the Earth germinates a seed, and as well humans emerge from a woman's vulva. These similarities led the Inkas to as well think that the first of the Ayllu came from a cavern or a hole (T'oqo).

The T'oqo of the firmament is in a dark area, called Lajha Haykuna. The spring or lagoon of the firmament is the Inka constellation of Qollqa, otherwise known as the Pleiades. All other celestial objects were thought to have come from these sources.

The elements of Quechua language and culture that were described in this chapter will greatly facilitate a more accurate sense of the cosmovision and astronomy of the Inkas.

5

Archaeoastronomy

Contents

Celestial Sphere	59
Heavenly Motions	60
Mid-latitudes	60
Qosqo	61
Solstices	61
Equinoxes	61
Zenith and Anti-zenith Sun	62
Horizon Astronomy	63
Cultural Astronomy Tools	63
Horizon Deviation	64
Summary	66

Archaeoastronomy is sometimes said to be the anthropology of astronomy. Celestial orientations appear in constructions found in many societies, implying there was great ancient interest in the Sun, Moon, planets, and stars for interaction with daily life. This was more than just a means of telling time, though, as such orientations have been found with respect to many aspects of society and culture. This chapter is an overview of some astronomical concepts and understanding these scientific basics is also necessary to fully comprehend Inka astronomy and cosmovision.

Celestial Sphere

People throughout history have observed motions in the sky; it is their uses for them that differ. Ancient cultures studied such events closely and many became experts on what can be called "horizon astronomy." In the Southern

© The Author(s), under exclusive license to Springer Nature Switzerland AG 2024
S. Gullberg, M. Rojas Gamarra, *Inca Cosmovision*, Astronomers' Universe, https://doi.org/10.1007/978-3-031-67580-5_5

Hemisphere the Earth's apparent motion is from left to right, or clockwise, when looking north toward the equator. As a result, the stars, Sun, Moon, and planets all rise from the east to the right and set to the left in the west. They appear to cross a celestial sphere in their daily motions (Gullberg, 2020).

Heavenly Motions

Ancient observers in cultures throughout the world saw different things in the sky. This gave rise to various cosmologies as they were influenced by the orientation of the view from their specific locations (Urton, 1981). People typically view astronomical occurrences from their own perspectives and therefore need to make a conscious effort to see the sky from the eyes of others (Aveni, 1981b; Gullberg, 2020).

Mid-latitudes

Qosqo is in the mid-latitudes, so bodies viewed there rise in the east, crossed the sky in an arc, and set on the western horizon. In the Southern Hemisphere looking north these bodies travel right to left, and in the Northern Hemisphere looking south from left to right. The altitude of the arc of travel varies throughout the year due to the tilt of the Earth's axis and to what degree is dependent upon the latitude of the observer as well as the position of the Earth in its orbit about the Sun. The annual variations of the elevation of this arc are responsible for daily changes in the horizon positions for sunrise and sunset (Gullberg, 2020).

Stars, as well as the Sun and Moon, travel east to west on their own respective arcs. In the Southern Hemisphere, when looking to the south, stars that are near to the south celestial pole travel in circles around the pole due to the Earth's rotation about its axis. These stars are called *circumpolar* and how many can be viewed always above the horizon is dependent upon the latitude of the observer. As viewed from the latitude of Qosqo, these would be the stars whose circles' lower limbs lie within 13° of the southern horizon (Gullberg, 2020).

Qosqo

Qosqo lies in the Andes at 13.532° south latitude. In Qosqo, when facing north the Sun and other ecliptic bodies will rise in the east on the right, cross the sky in arcs to the north, and set on the left in the west. The region surrounding Qosqo experiences a Sun in June that rises in the northeast, crosses the sky on a relatively lower arc, and correspondingly sets in the northwest. In December the solstice Sun rises in the southeast, follows a higher path, and sets in the southwest. The Inkas and the civilizations from which they learned observed the repetitive and predictable patterns of the horizon positions of sunrises and sunsets. They used this for calendrical systems for crop management and religious festivals. The Inkas built solar pillars on the horizons of Qosqo to predict and mark the positions of solar events (Gullberg, 2020).

Solstices

When the Sun reaches its southernmost declination of 23.439° on its apparent ecliptic path of travel we observe what is called the December solstice. In the Southern Hemisphere this is also known as the summer solstice and throughout the months leading up to it there are small southward movements of the position of the sunrise on the horizon each day, except for about 2 days before and after the solstice when there is no observable motion as the Sun "stands still" before reversing course back to the north (Bauer and Dearborn 1995; Urton, 1981). The Sun's northernmost declination occurs at the June, or winter, solstice when similar motions are observed, only in reverse. The Sun travels across the sky on a lower arc in June but crosses high in the sky at the time of the solstice in December (Gullberg, 2020).

The Inkas learned the cycles of solstices and equinoxes and used this for their annual crop management. Zuidema (1981) and Aveni (1981a) described solar pillars that no longer exist on the horizon of Qosqo. Their existence was recorded by chroniclers following the Spanish conquest (Gullberg, 2020).

Equinoxes

Similar to solstices, there are two solar equinoxes. The March and September equinoxes occur when the Sun crosses the celestial equator, from south to north in March and from north to south in September. In the Southern

Hemisphere the September and March equinoxes signal the beginnings of spring and fall respectively (Gullberg, 2020).

The precise moment that the Sun crosses the celestial equator is called an equinox. This occurs twice each year on or about March 21st and September 22nd and on those dates, there are equal periods of sunlight and darkness. The longest days and nights occur each year when the Sun reaches the southernmost and northernmost points in its ecliptic travel (23.439° south or north) on or about December 21st and June 21st. These are the December and June solstices, and in the Southern Hemisphere are commonly known as the summer and winter solstices (Gullberg, 2020).

Andean traditions are said to have linked equinoxes with agriculture, September in the spring when the soil was first prepared for planting and in March in the fall when maize was harvested. Bauer and Dearborn (1995) state that the festival of the Sun called Citua Raymi was celebrated every September at the equinox, and perhaps Paqa Puqay Quilla at the equinox in March (Zuidema, 2007). Inka awareness of the equinoxes is mentioned in Spanish chronicles (de la Vega, 1961[1609]), but conclusive evidence is elusive (Dearborn & White, 1989; Gullberg, 2020).

Zenith and Anti-zenith Sun

As the Sun travels along the ecliptic, twice each year it passes directly overhead any location lying between the Tropic of Cancer and the Tropic of Capricorn. All sites between the latitudes of N23.439° and S23.439° have two times without shadow when the Sun passes directly overhead at local noon. At sites precisely on the tropic latitudes, though, the Sun is directly overhead only 1 day per year at the times of the solstices (Gullberg, 2020).

Specific dates for the zenith Sun also depend on the latitude of the observer. In Qosqo, the two dates for zenith passage occur on February 13th and on October 30th (Zuidema, 1981). Zenith passage is observed when a vertical gnomon casts no shadow at local noon, and on those dates the Inkas made note that the Sun rose on the horizon at approximately 103.5° and set at about 256.5°. Zuidema (1981) suggested that the Incas also observed days of anti-zenith, or nadir, passage of the Sun, and stated:

> The only way in which the Incas could have determined the dates of August 18 and April 26 would be to note the days when the sun goes through the zenith. Reversing the direction of sunrise on those days would have given them the days of antizenith (= nadir) sunset. (p. 322)

Anti-zenith passage takes place in Qosqo on April 26th or 27th and August 18th or 19th, with sunrise on these days occurring at 076.5° and sunset at 283.5°. These dates closely correlate with Inka festivals of maize planting and harvest, which supports interest for such observances (Zuidema, 1981; Gullberg, 2020).

Horizon Astronomy

As the Earth revolves around the Sun the tilt of its axis is what causes the horizon positions of sunrises and sunsets to move slightly every day. Certain natural features were used by the Inkas to mark the positions of the Sun on days of solstices and equinoxes. Alignments of buildings and shrines could also have been constructed to assist in identifying these important dates; the Inkas erected solar pillars on Qosqo's horizons to help make these observations. Others such pillars were erected at Q'espiwanka in the Sacred Valley (Bauer and Dearborn 1995; Gullberg, 2020; see Figs. 5.1, 5.2, and 5.3).

In the location of Qosqo, the June solstice sunrise occurs at approximately 064° and June solstice sunset at 294°. At December solstice the Sun rises at 114° and sets at 244°. Numerous extant waqas exhibit orientations for the solstices. *The Huarochirí Manuscript* by Salomon and Urioste (1991) states that the Inkas had professional observers called *Yancas* who made these astronomical observations. Yancas observed the course of the Sun and its alignment with "calibrated walls" from designated vantage points (Salomon & Urioste, 1991; Gullberg, 2020).

Cultural Astronomy Tools

Tools commonly used for archaeoastronomy research include theodolites, GPS receivers, sighting compasses, inclinometers, digital cameras, tape measures, computers, and astronomical software. The theodolite, or surveyor's transit, measures angles very precisely, to the arcsecond, and is used to verify alignments that require the greatest accuracy (see Fig. 5.4). GPS receivers give coordinates of latitude and longitude, as well as elevation above sea level. A professional sighting compass is useful for measuring azimuths and alignments in far less time than with a theodolite. An inclinometer is used to measure inclination above or below the level horizon (see Fig. 5.5). Digital photographs are essential for documentation of research, and especially in recording actual evidence of effects of light and shadow. Tape measures are

Fig. 5.1 The eastern pillar above Q'espiwanka

used for dimensions of certain object(s). Computers and astronomical software such as *Stellarium* provide insight into past skies that were viewed by ancient societies and also can be used to predict alignments and celestial events. *Google Earth* is a valuable tool as well. This program makes available world-wide satellite imagery in formats that allow preliminary evaluation of sites before going to the field.

Horizon Deviation

Horizon deviation occurs when the visible horizon differs from the astronomical horizon in areas of mountainous terrain and is especially a factor at latitudes where the arc of travel for a specific body is angled such that sunrise or sunset over the mountains will be at a measurably different point than what is given for the astronomical horizon (Gullberg, 2020).

Fig. 5.2 The pillars above Q'espiwanka on the Cerro Sayhua ridge

Fig. 5.3 Sunrise over the right Cerro Sayhua pillar as viewed from Q'espiwanka's central granite boulder in June just prior to the solstice; the sunrise each day will continue to move toward the left pillar as the solstice approaches

Fig. 5.4 Theodolite at Q'espiwanka. (Watercolor printed with permission. © 2024, Jessica Gullberg. All rights reserved)

When the inclination (angular altitude) of the horizon above level is known by measurement with an inclinometer, the latitude is obtained from a chart or by GPS measurement, and the declination of the Sun is taken from the nautical almanac, the following formula of spherical trigonometry can be used to calculate the actual position that the Sun will rise on the horizon:

Azimuth of Sun = Arccos((Sin(DEC) — sin (LAT)Sin(ALT))/ (Cos(LAT)Cos(ALT))

And, of course, there exists software that will do this calculation for you when you enter the numbers.

Adjustments for horizon deviation are frequently needed in archaeoastronomical research. Examples of the degree of deviation along the horizon are shown in Table 5.1 (Gullberg, 2020).

Summary

Paths of heavenly bodies crossing the sky vary with latitude, and throughout Qosqo, the Sacred Valley, and Machu Piqchu motions observed are typical of those found in much of the tropics.

Fig. 5.5 Suunto Tandem Sighting Compass and Inclinometer

Table 5.1 Horizon azimuth shift for altitude

Altitude above astronomical horizon	Latitude 13.5°	Latitude 40°
1°	0.22°	1.04°
2°	0.49°	1.99°
4°	1.07°	3.84°
6°	1.69°	5.63°
8°	2.35°	7.37°
10°	3.04°	8.06°

Solstices and equinoxes were of great interest to the Inkas, and they used these as a calendar to regulate crop management and the dates of religious festivals. The Inkas built solar pillars to assist them in recognizing these key astronomical events.

Horizon astronomy is the observation of risings and settings of the Sun and other bodies on the horizon. Many ancient civilizations, including the Inkas, practiced horizon astronomy to manage societal functions.

Various instruments are used for archaeoastronomical research. These include theodolites, GPS receivers, sighting compasses, inclinometers, digital cameras, tape measures, computers, and astronomical software. Mathematical formulae and/or computer software may be used to accurately depict, predict, or verify the horizon positions of various astronomical events. Researchers must be aware of the effects of horizon deviation, a primary concern for solar research in mountaneous areas such as in the Coordillera de los Andes.

6

Facets of Inka Astronomy and Cosmology

Contents

Qorikancha	70
Solar Worship	70
Cosmological Origins	71
Festivals	71
Alignments in Architecture	72
Ushnus	73
Sukanqas	73
Pillars	75
The Milky Way	75
Orientation and Quadripartition	75
Celestial River	76
Dark Constellations	77
Star Constellations	77
Stars	77
Planets	78
The Pleiades	78
Seq'e System and the Stars	79
Q'enqo Grande	79
Lacco	82
Waqa 44	84
Q'espiwanka	85
Machu Piqchu Region	86
Summary	92

The original version of the chapter has been revised. A correction to this chapter can be found at https://doi.org/10.1007/978-3-031-67580-5_12

© The Author(s), under exclusive license to Springer Nature Switzerland AG 2024, corrected publication 2025
S. Gullberg, M. Rojas Gamarra, *Inca Cosmovision*, Astronomers' Universe, https://doi.org/10.1007/978-3-031-67580-5_6

The Inkas venerated the Sun, and their emperor was said to be "the son of the Sun." They also were aware of many stars and planets and paid particular attention to the Milky Way and the Pleiades. Horizon astronomy was used to mark the passage of sunrises and sunsets on their horizons in order to manage time for agriculture and religion.

Pachakuteq Inka Yupanqui was the Inkas' ninth ruler and he succeeded his father, Wiraqocha Inka, following a legendary battle with the Chanca in 1438. This pivotal conflict took place roughly two centuries after the founding Inkas first entered the Qosqo valley. Nearly 100 years later the Spaniards began their conquest of the Inkas which brought any further astronomical development to a halt (Aveni, 1981a; Gullberg, 2020).

Inka astronomical knowledge had origins long before Pachakuteq. The Inkas learned from what had been collected for centuries and added additional information from cultures they assimilated, such as the Huari, Nasca, and Chavin (Aveni, 1981a). As a result, the Inkas developed a very good understanding of astronomy. Astronomy in Qosqo was one with religion and everyday life, and the Inkas believed what they saw in the heavens was directly connected with activity on the Earth (Urton, 1981). Successful promulgation of State programs to promote the legitimacy of solar worship required a detailed knowledge of the movement of the Sun. Prosperous crop management depended on knowing the best times for planting and harvest and the Inkas managed this by using solar horizon positions. This chapter outlines facets of Inka astronomy and cosmology that will help to better understand the oral traditions in chapters to follow (Gullberg, 2020).

Qorikancha

The Inkas' primary Temple of the Sun was the Qorikancha of Qosqo (Fig. 6.1). The Qorikancha's west wall is aligned approximately for the horizon point of the heliacal rise of the Pleiades in June (Zuidema, 1982). Additionally, the December solstice sunset could be seen from the Qorikancha with two pillars on the horizon at ChinchInkalla (Aveni, 1981a, b; Gullberg, 2020).

The Inkas knew the Sun's annual cycle and its path along the horizon. They used this to determine times to plant and harvest (Hemming & Ranney, 1982; Gullberg, 2020).

Solar Worship

The Inkas believed their emperor to be the son of the Sun. The Sun was also venerated for its life-giving role in agriculture. Rituals and ceremonial travel to pilgrimage centers were developed to support this ideology (Bauer & Stanish, 2001). Pachakuteq required that the Inkas worship the Sun and

6 Facets of Inka Astronomy and Cosmology

Fig. 6.1 The author Steven and his son, Steven II, at the Qorikancha—the Spaniards built a cathedral on top of the original Inka site. Published with permission by Steven Gullberg

related temples were built throughout the Tawantinsuyo. The calendar was used to know when to conduct annual religious and agricultural festivals. Spanish chroniclers wrote that solar pillars were built on Qosqo's horizons for such calendrical purposes (Zuidema, 1981; Aveni, 1981a, b; Gullberg, 2020).

Cosmological Origins

The Inkas believed the Sun and Moon originated from the *Isla del Sol* and the *Isla del Luna* on Lake Titicaca. It was said that the Sun rose from an outcrop of rocks on the island that now bears its name and the Moon emerged from a smaller island nearby (Gullberg, 2020).

Festivals

Intirraymi, the festival of the Sun at the time of the June solstice, was perhaps the most significant celebration. These Inka ceremonies lasted for 8 days and were attended by the royal mummies; they included sacrifices and great chanting (Hemming & Ranney, 1982). The next greatest time of solar worship was at December solstice with the festival of Qhapaq Raymi (Dearborn & Schreiber, 1986; Gullberg, 2020).

Alignments in Architecture

Structures were designed to assist Inka astronomy and cosmology. Observation of the heliacal rise of the Pleiades was important and architectural alignments could serve to help guide the viewer's eyes to the proper point on the horizon. Zuidema and Aveni examined the Qorikancha of Qosqo and found it to be oriented this way (Fig. 6.2). They found an alignment for the Pleiades that

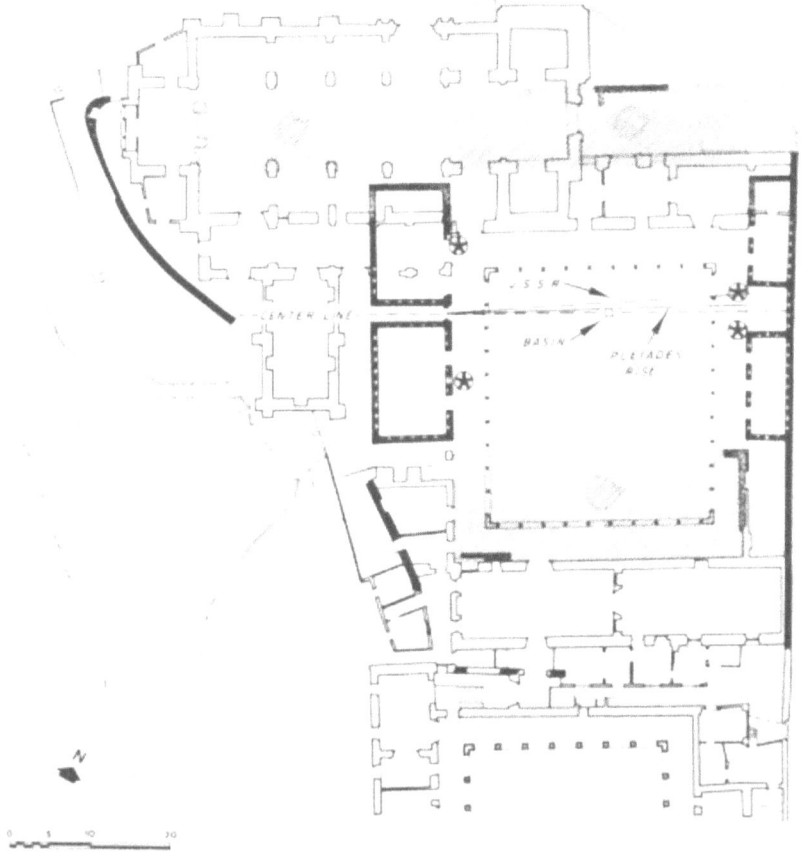

Fig. 6.2 Plan of the Qorikancha. The sightline for the heliacal rise of the Pleiades passes between two western rooms, across a ceremonial basin and between two eastern rooms. The two encircled stars indicate positions that had been used for gold or precious stones. (Reprinted with permission from (Zuidema, 1982). © 2006, John Wiley & Sons. All rights reserved)

passed between two western rooms, across a ceremonial basin, between holes for gold and precious stones, and then through a space between two eastern rooms. The Qorikancha was a walled enclosure of stone structures with thatched roofs (Salazar & Salazar, 2014). Zuidema and Aveni determined that the western and eastern walls of the Qorikancha were aligned for the Pleiades at an azimuth of 66° 44′. They found that the Pleiades rose at 65° 58′ in 1500 CE; 66° 22′ in 1400 CE; and 66° 46′ in 1300 CE (Zuidema, 1982). However, in more recent research Ziolkowski and Kosciuk (2018) found little to support precise astronomical observation at the Qorikancha. They do support precise observations in one small window, but dispute much of the earlier work (Gullberg, 2020).

Ushnus

Ushnus were platforms for making offerings, for observing visual sightlines, and for receiving liquid offerings for the Pachamama, such as chicha. The Inka also sat on top of the ushnu to dispense justice (Zuidema & Quispe, 1973). Some ushnus were waqas and some were not—those that weren't were not prayed to or asked for advice and guidance. "The Inkas have places called *usnos* designated throughout this kingdom for sacrifice, which is always for the *capacocha* to the Sun and the *waqas* …. The Inka made sacrifices to his father the Sun with gold, silver, handsome ten-year old boys and girls who had no blemishes, not even a mole" (Guaman Poma de Ayala, 2009 [1615], p. 201; Gullberg, 2020).

Waqas were venerated animate or inanimate objects and could be many things such as natural features of the landscape, trees, springs, and rocks. Waqas were thought to be possessed by a local deity whose life force gave prosperity to all around it (Staller, 2008). The waqas were worshipped and received offerings (Gullberg, 2020).

While an ushnu could give divine power to those seated on it, it did not emanate prosperity. Ushnus were platforms with several tiers and a staircase on one side (Gamarra et al., 2024; Staller, 2008; Guaman Poma de Ayala, 2009 [1615]; Gullberg, 2020; see Fig. 6.3).

Sukanqas

The pillars of Pucuy Sukanqa were positioned to mark the horizon position of the Sun at the start of the rainy season and the pillars of Chirao Sukanqa indicated the beginning of the dry season. Other pillars were erected to establish the time for festivals and the passage of months. Pucuy Sukanqa and Chirao

Fig. 6.3 Drawing of an Ushnu. (Reprinted with permission from (Guaman Poma, 2010). © 2010, University Texas Press. All rights reserved)

Sukanqa were each waqas on the Qosqo seq'e system (Zuidema, 2005; Gullberg, 2020).

Sukanqas, by definition, marked sunrise or sunset points and the pillars of Pucuy and Chirao helped to define the zenith sunrise and the anti-zenith sunset. Sayhuas were markers as well and those near Qosqo used the position of sunrise to indicate the start of Inka months. Zuidema (1981) states the system included 12 sayhuas and two sukanqas. The Chirao Sukanqa was likely located on Cerro Piqchu to the northwest of Qosqo and the Pucuy Sukanqa southeast on a mountain at Quispicancha (Tipon). Zuidema continued that the names *Pucuy* and *Chirao* were for seasons—Pucuy for "the time when fruits ripen during the latter part of the rainy season" and Chirao for "the time when the first changes in the weather occur after the dry season" (p. 330). Solar alignment with the Pucuy Sukanqa was for a date in February at the time of zenith sunrise, and the Chirao Sukanqa aligned for a date in August and the anti-zenith sunset (Gullberg, 2020).

Pillars

Cobo (1983 [1653]) described solar pillars on the horizon of Qosqo to track solar movement. These no longer remain, but two such pillars are located at 3860 masl on the Cerro Sayhua ridge above Q'espiwanka, the palace of Wayna Qapaq, and are aligned for the June solstice sunrise when viewed from a large boulder located in the center of the palace courtyard (see Figs. 5.1, 5.2, and 5.3). None of the pillars that Spanish chroniclers described at Qosqo survived the purge of idolatry, but these two towers near the present-day community of Urubamba show that such solar pillars likely did exist (Gullberg, 2020).

The Milky Way

Inka cosmology saw the Milky Way as a river flowing across the night sky in a very literal sense. Inkas envisioned earthy waters drawn into the heavens and later returned to Earth after a celestial rejuvenation. It was thought that the Earth floated in a cosmic ocean and when the "celestial river" dipped into that ocean its waters were drawn into the sky. "The Milky Way is therefore an integral part of the continuing recycling of water throughout the Quechua universe" (Urton, 1981, p. 60; Gullberg, 2020).

Orientation and Quadripartition

The Milky Way passes brightly overhead in Qosqo and the Inkas observed it closely (Urton, 1981). They saw the galaxy as two separate rivers because of the Milky Way's alternating position on the horizon each 12 h due to the Earth's rotation. The plane of the Milky Way is inclined 26° and 30° with the Earth's axis. The orientation is 26° degrees toward the south celestial pole and 30° toward the north (Urton, 1981). The Milky Way at times can be seen rising in the southeast, passing through the zenith, and setting in the northwest. Twelve hours later the horizon positions have shifted, and the galaxy instead rises in the northeast, passes through zenith, and sets in the southwest. This

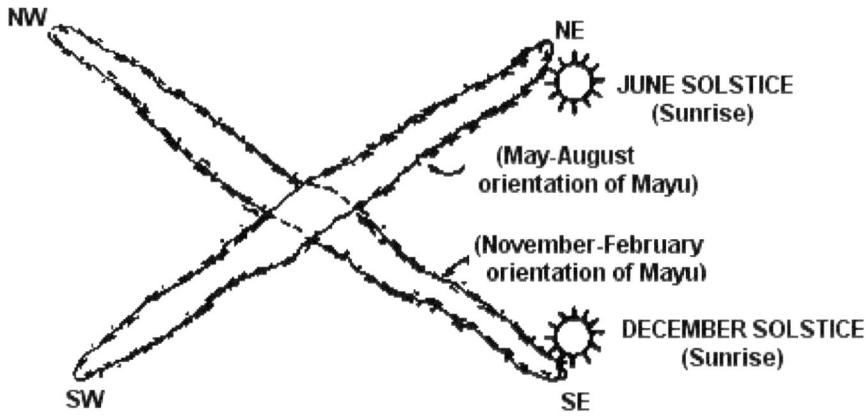

Fig. 6.4 Inka Quadripartition of the Sky. (Reprinted with permission from (Urton, 1981). © 1988, University Texas Press. All rights reserved)

24-h cycle creates two zenith-intersecting intercardinal axes that divide the celestial sphere into four quarters (Urton, 1981; Gullberg, 2020; Fig. 6.4).

At the time of the December solstice, when the Sun rises at 114° on Qosqo's horizon, the evening positioning of the Milky Way aligns with it. During the June solstice sunrise at 064° the Milky Way aligns in a similar manner. The Sun rises and travels with the Milky Way only at times of solstices (Urton, 1981). The Inkas ordered their sky with this celestial quadripartition instead of the general direction of ecliptic travel. Urton (1981) said that this gave them a nearly 90° difference in their orientation to the heavens when compared with that of Western astronomy (Gullberg, 2020).

Celestial River

Inka cosmology links the Milky Way with the Willqamayu (Vilcanota River). The Willqamayu flows southeast to northwest through the Sacred Valley, past Machu Piqchu and beyond. Its was these waters that were thought to rise into the Milky Way and later fall to Earth as rain. The Inkas believed the Sun was stronger in the summer because it drank from a swollen Willqamayu while traveling beneath it during the night. In the winter it was weaker because it has had less to drink. The Milky Way was thought to be a heavenly reflection of the Willqamayu (Urton, 1981; Gullberg, 2020).

Dark Constellations

The Inkas envisioned *dark constellations*, the shapes of beings formed in dark sections in the visible band of the Milky Way. These dark patches are formed by interstellar gas and dust blocking light from the stars beyond. The Inkas saw great cosmological characters that were though to be responsible for the procreation and augmentation of their corresponding creatures on Earth (Gullberg et al., 2020).

The dark constellations are most visible at the time of the March equinox when they can be seen in the Milky Way at midnight. During the September equinox the fewest are visible because they are beneath the horizon for much of the night. At the solstices the Milky Way briefly aligns with the horizon points of the respective sunrises and sunsets and the dark constellations appear to follow its path (Urton, 1981). Urton said that the dark constellations are related to the rainy season and thus "it is … essential to study the Dark Cloud constellations by analyzing the connections between sky and water and earth and water" (Urton, 1981, p. 173; Gullberg, 2020; see Figs. 9.4, and 9.5).

Star Constellations

In addition to dark constellations, the Inkas also recognized certain star to star groupings. Inka star constellations are located in close proximity to the Milky Way, especially between Taurus and Orion. While dark constellations primarily represented animals, star to star constellations were more often associated with objects (Urton, 1981). Cobo (1990 [1653]) wrote that the Inkas designated stars as patrons of animal species, and that those stars were venerated for that reason (Gullberg, 2020).

Stars

The Inkas looked at the heavens and imagined cosmological shapes with meaning for them. They did not always discern between planets and the brightest stars, however. The Quechua word *Chasca* was used for both, while *Coyllur* was used for stars of lesser prominence (Bauer and Dearborn 1995). The Inkas believed that these celestial counterparts also had influence over the health and reproduction of those below (Gullberg, 2020).

Planets

Inka planetary observation is not well documented; Venus, however, was seen both as an evening star and as a morning star. It can sometimes be seen in the daytime, but it does not cross the sky at night. Bauer and Dearborn (1995) relate that Venus had many names such as *Chasca Cuyor, Pacaric Chasca, Pacari Cuylor, Auquila, Pachahuárac, Chachaquaras, and Atungara* (Gullberg, 2020).

The Pleiades

The Pleiades, or *Qollqa*, were of great importance in Inka astronomy because the Inkas believed that all water constellations emanated from there and also found Qollqa to be useful for predicting crop success (see Fig. 9.1). The Spanish chronicler Cobo (1990 [1653]) said that they were called *Collca* and that the "power that conserved the animals and birds flowed from this group of stars" (p. 30). Qollqa disappears behind the Sun for approximately 37 days and first returns to view around June ninth, about 12 days before the solstice sunrise. Even today the brilliance of these stars is assessed by Andean farmers with a bright appearance indicating a season of rain with a good harvest. Coversely, a dull appearance resulting from obscuration by high cirrus clouds indicated drought. This method had been discovered to anticipate the arrival of El Niño. Orlove, Chiang, and Cane (2000) relate: "…find that poor visibility of the Pleiades in June – caused by an increase in subvisual high cirrus clouds – is indicative of an El Niño year, which is usually linked to reduced rainfall during the growing season several months later" (p. 68). Sightings were likely made over several days during the approach of Intirraymi on the 24th (Orlove et al., 2000). With a prediction of drought, farmers delayed planting potatoes because they are especially susceptible to drought (Gullberg, 2020).

The Inkas worshipped Qollqa and thought of it as a predictor of life and death. Its brilliance determined when maize should be planted and how plentiful a crop to expect. The exact date of the heliacal rise is somewhat a function of the Moon's phase and atmospheric conditions (Bauer and Dearborn 1995). The Pleiades figured prominently with several ancient cultures but were especially important in the Andes because of their use in predicting significant climatic changes during El Niño years (Gullberg, 2020).

Several structures exist with possible alignments related to Qollqa's heliacal rise. The Qorikancha of Qosqo and the Sun Temple at Llaqtapata both have orientations for June solstice sunrise and the heliacal rise of Qollca. Corridors at each could have been used to guide observer's eyes to the point on the horizon where Qollqa would first appear. Dearborn and Schreiber (1986) also determined that the window in what is called The Torreon at Machu Piqchu is aimed for these risings as well (Gullberg, 2020; see Fig. 3.1).

Seq'e System and the Stars

The Inkas' seq'e system surrounding Qosqo was extensive. The 41 seq'es were filled with waqas for worship and care, at least 328 that were maintained by panaqas and ayllus as part of their state-assigned responsibilities. Zuidema said that some of the seq'es were straight and included intentional orientations for the rising and setting of certain stars. Among these were seq'e alignments for stars known as Betelgeuse, the Pleiades, Vega, and β Centauri (Zuidema, 1977, 1982, 1990; Gullberg, 2020).

Q'enqo Grande

Examples such as the "Awakening of the Puma" at Q'enqo Grande demonstrate a great knowledge of horizon astronomy and also the degree of creativity the Inkas were capable of when constructing their waqas. On the top of this large outcrop they carved two gnomons that form the sacred puma's eyes and with them created a fissure in a nearby low wall that allows light to fall on these gnomons during the June solstice sunrise in such a way that creayes a "puma" visual effect of light and shadow (Gullberg, 2020; see Figs. 6.5, 6.6, and 6.7).

The chamber within Q'enqo Grande includes caves, niches, and ritual steps. The primary altar is finely carved and might well have been used for ceremonies. The light tube in the cave's upper northwest corner admits light that could be reflected from gold plates to illuminate the rest of the chamber. The cave includes a special effect of light climbing three ritual steps (not shown) below the light tube adjacent to the far side of the altar in Fig. 6.8 (Gullberg, 2020).

Fig. 6.5 Q'enqo Grande and its monolith

Fig. 6.6 Fissure aligned for the June solstice sunrise

Fig. 6.7 The Awakening of the Puma at June solstice sunrise

Fig. 6.8 Q'enqo Grande's primary altar

Lacco

Lacco is another large outcrop that includes several examples of the passion the Inkas had for using light from the Sun or Moon in their waqas and ceremonies. Its three caves show the prowess the Inkas had for solar orientations. Each has an altar that is illuminated at certain times by the Sun or Moon. The altar within Lacco's northeast cave altar is illuminated in the early morning on the days surrounding the June solstice and the cave's opening is centered on the solstice sunrise on the horizon. The Sun's rays are cast directly on the altar and reflections of that light illuminate the rest of the cave (Gullberg, 2020; see Figs. 6.9, and 6.10).

Lacco's southeast and southwest caves use specifically oriented light tubes to admit light from the Sun or the Moon. The southwestern cave is the smaller of the two and also has a smaller altar (see Fig. 6.11). The cave to the southeast, known as the Temple of the Moon, was the most prominent of the three as evidenced by the degree of workmanship in its sculpture, complete with fine carvings of a puma and snake near its entrance and the highly polished

Fig. 6.9 June solstice sunrise as seen from Lacco's Northeast cave. The Sun advances each day from the right until reaching the central standstill as shown, after which it reverses course and moves back to the right again

Fig. 6.10 Lacco's Northeast cave and altar illuminated by the June solstice early morning Sun.

Fig. 6.11 The crescent Moon as viewed upward through the light-tube in Lacco's southwestern cave

Fig. 6.12 The Sun as viewed through Lacco's southeast cave's light-tube

altar within its inner chamber. The altar is illuminated by the Sun when it is 90° above the horizon near the time of zenith passage. Considering the cave's given name, lunar ceremonies would also have been conducted here as well (Gullberg, 2020; see Figs. 6.12, 6.13, and 6.14).

Waqa 44

Using the principles of Ayni and Kawsaypacha, Amauta Emilio Huaman Huillca discovered the significance of Waka 44 in 1975. Because this is a waqa (and waqas have always been places of peoples' pilgrimages for ceremonies or to ask for something), our Amauta went there many times, both to visit and also to observe; he wondered about the purpose of this waqa. He said that he dreamed about this and in a vigil state he was able to visualize the guidelines of tangent lines between the two cylinders carved in stone and also the line that crosses them both. Projections of these lines to the horizon indicate horizon positions for sunrise and sunset at the June and December solstices and as well as those for the March and September equinoxes (see Fig. 6.16). Since

Fig. 6.13 The southeast cave's altar illuminated by the zenith Sun.

1990 author Milton has gone there with Amauta Emilio many times to examine Waqa 44. Author Steven later examined the Waqa 44 alignments independently before he became aware of the work of Milton and Amauta Emilio (see Figs. 6.15, 6.17, 6.18, and 6.19). This supports what Amauta Emilio found.

Q'espiwanka

The pillars above Q'espiwanka are important examples of Inka history. Spanish chroniclers recorded solar pillars on the horizons of Qosqo, but all were eradicated in the Catholic extirpation of Inka idolatries. Accounts in the chronicals describe pillars being used to monitor solar horizon positions, but no evidence of this remains near Qosqo. Pillars near present-day Urubamba, however, give credibility to the reports because two towers exist on the Cerro Sayhua ridge and are oriented for June solsice sunrise. This can be seen at the time of June solstice sunrise when viewed from a sacred granite boulder that was central on Wayna Qhapaq's palace grounds (Gullberg, 2020; see Figs. 5.1, 5.2, and 5.3).

Fig. 6.14 Author Steven at the altar in Lacco's southeast cave. (Watercolor printed with permission. © 2024, Jessica Gullberg. All rights reserved)

Machu Piqchu Region

Machu Piqchu (Machu Picchu) has many astronomical orientations and one of the most fascinating is the axis of the June solstice sunrise and December solstice sunset. Machu Piqchu's Sacred Plaza is located on what could have been a seq'e that included the River Intiwatana in the gorge below and the Sun Temple of Llaqtapata on a distant ridge. This orientation is clearly demonstrated when sunrise over Machu Piqchu is viewed from Llaqtapata on the morning of the June solstice. Polo de Ondegardo (1965) [1571]) said that

Fig. 6.15 Huaca 44 and its two cylinders that are aligned to indicate solar horizon events

Fig. 6.16 The cylinders are aligned to indicate the directions of the horizon positions of sunrise and sunset on days of the solstices and equinoxes; JSSR—June solstice sunrise; ESR—Equinox sunrise; DSSR—December solstice sunrise; DSSS—December solstice sunset; ESS—Equinox sunset; JSSS—June solstice sunset

Fig. 6.17 Large and small cylinders measured by author Steven's son, Greg, are carved into the waqa's base. Published with permission by Greg Gullberg

6 Facets of Inka Astronomy and Cosmology 89

Fig. 6.18 Measurement for the June solstice sunrise orientation

Fig. 6.19 June solstice sunrise as viewed between the cylinders (author Steven's photograph)

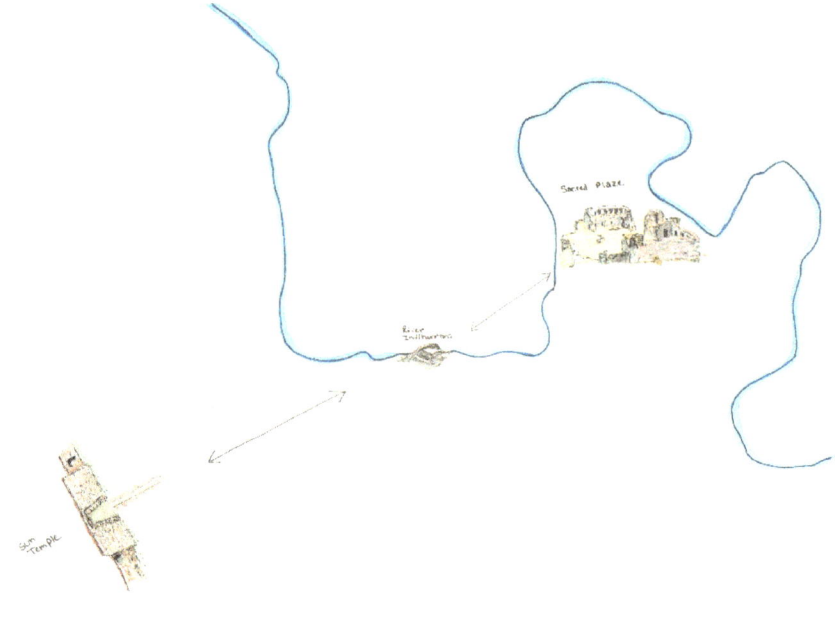

Fig. 6.20 Relative positions (left to right) of the Llaqtapata Sun Temple, the River Intihuatana, and the Machu Piqchu Sacred Plaza on the axis of the June solstice sunrise and December solstice sunset. (Watercolor printed with permission. © 2024, Jessica Gullberg. All rights reserved)

each Inka Llaqta (village) had seq'es connecting waqas and here it appears that a seq'e system may have been established at Machu Piqchu. The potential existence of seq'es and waqas other than in Qosqo is intriguing (Gullberg, 2020; see Figs. 6.20 and 6.21).

It is now thought that the River Intiwatana (see Fig. 6.22) may have played a significant role in a greater ceremonial complex surrounding Machu Piqchu. It provides a distinct link between the waqas and structures of Machu Piqchu and those at Llaqtapata and may be part of this seq'e system connecting them. This alignment and potential seq'e indicate the ceremonial significance of the River Intiwatana. Llaqtapata's Sun Temple could have been positioned where it is as part of this alignment. The orientation of the Sun Temple with Machu Piqchu's Sacred Plaza on the axis of June solstice sunrise and December solstice sunset is fascinating. The rediscovery of Llaqtapata and its many structures, all engulfed in the cloud forest, suggests that elaborate ceremonies for pilgrimage may have taken place throughout the area. A greater

6 Facets of Inka Astronomy and Cosmology   91

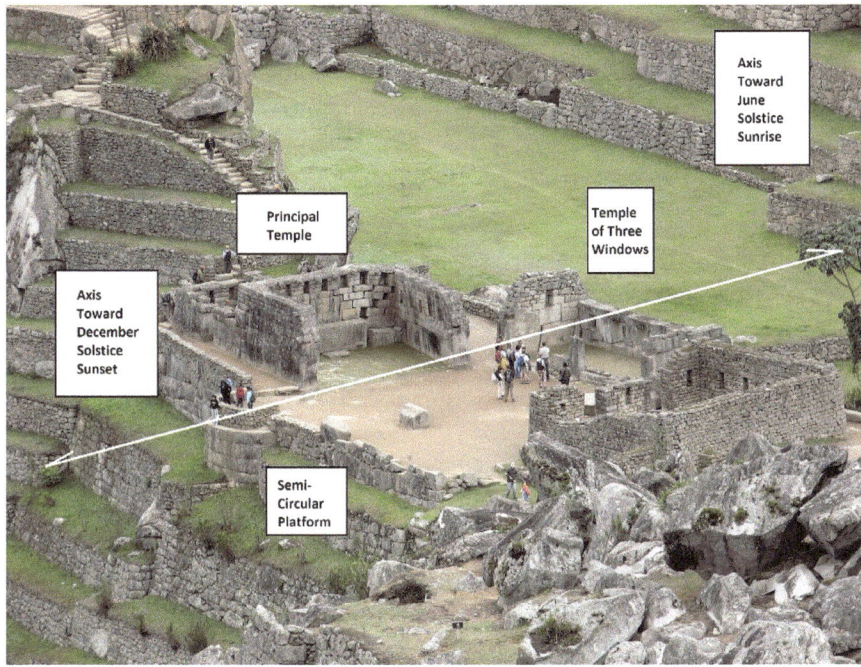

Fig. 6.21 Machu Piqchu's Sacred Plaza orientation

Fig. 6.22 The Intiwatana stone in the River Intiwatana complex

Fig. 6.23 The stone-lined channel leading from the Llaqtapata Sun Temple's central doorway

understanding of the River Intiwatana emerged after Llaqtapata, only 5 km from Machu Piqchu, was located in 2003 (Gullberg, 2020).

Research at Llaqtapata has given rise to many questions regarding the overall extent, orientation, and function of the entire Machu Piqchu region. The orientation of the stone-lined channel directed from Sun Temple at Llaqtapata over the River Intiwatana and to Machu Piqchu's Sacred Plaza is striking and can be clearly seen from Llaqtapata at sunrise on the day of the June solstice. An orientation for the heliacal rise of the Pleiades also exists at Llaqtapata. It has been hypothesized that the Inkas may have poured ceremonial liquids into the stone-lined channel as part of reenergizing the winter Sun (Gullberg, 2020; see Figs. 6.23, 6.24, and 6.25).

Summary

Extant Inka waqas and temples clearly demonstrate the sophisticated celestial knowledge that developed in the Andes. Inka cosmology began with worship of the Sun, and to a lesser extent the Moon, and with the assertion that the

Fig. 6.24 View from the primary door of Llaqtapata's Sun Temple with the stone-lined channel in the foreground aimed for Machu Piqchu's Sacred Plaza. (Watercolor printed with permission. © 2024, Jessica Gullberg. All rights reserved)

ruling Inka was the "son of the Sun." It was only natural that the Inkas made solar observations and developed astronomical models to reflect this.

Their interests went beyond the Sun, however, as they included other objects in the heavens. The Milky Way is very prominent at the latitude of Qosqo and the Andean peoples adopted it as their primary reference in the night sky. While European constellations were oriented with the ecliptic path of the Sun, the Inkas ordered their sky with the Milky Way.

The Moon played a significant role, but this is far less documented than the role of the Sun, likely due to the Moon's female connotations. The Inkas used both the Sun and the Moon with their calendar.

A calendar of Qosqo was visible on the horizon in the form of solar pillars on the city's horizon. These were built to track the annual path of sunrises and sunsets. The pillars of Qosqo were destroyed in the Spanish purge of idolatry, but two towers escaped and exist in the Sacred Valley above the present-day village of Urubamba.

Zuidema proposed an astronomically based calendar using 328 waqas on 41 seq'es surrounding Qosqo. The remaining 37 days of the 365-day tropical solar year were accounted for by the approximate period of the annual disappearance of Qollqa behind the Sun. Others have questioned his hypothesis.

Fig. 6.25 June solstice sunrise from Llaqtapata's Sun Temple. The water channel is below the camera tripod

Qollqa was worshipped by the Inkas and played a major role in agricultural production. By observing the relative brilliance of the heliacal rise of Qollqa, the Andeans had discovered a means of predicting the effects of El Niño. Orlove, Chiang, and Cane relate that the degree of brilliance and visibility of Qollqa predicted a good crop in normal years, or a depleted one with El Niño.

Much of Inka cosmology and myth is displayed in the system of dark constellations that were visualized in parts of the Milky Way where light was blocked by interstellar dust and gas. The world of Hananpacha above was closely intertwined with both the world of Kaypacha and the subterranean Ukhupacha. Powerful and influential spirits lived in many features of the natural worlds and also in the cosmos. The creatures of these dark constellations figured prominently in everyday life.

Inka astronomy was used throughout society and culture. The State created temples and waqas with astral orientations to give an aura of connectivity with the heavens to further establish power and legitimacy. The astronomical alignments that remain in waqas and temples today are testaments to the sophistication of the system that had evolved. The many celestial orientations found in the Inka empire can only be properly understood when taken in full cultural context and this will be further explored in the following chapters.

7

The Waqas of the Inkas

Contents
Summary 126

Waqas were a central feature in Inka culture and some had astronomical uses. Certain ones had astral orientations while others displayed intentional light and shadow effects. While most waqas were destroyed in the Catholic extirpation of idolatries, many that had been carved in rock remain (see Fig. 7.1). To truly learn Inka astronomy, you need to have a much greater understanding of waqas and their function in Inka culture, astronomically and otherwise. With help from what Garcilaso de la Vega and Guaman Poma de Ayala captured in their chronicals, as well as with author Milton's oral traditions, waqas are explored in great detail in this chapter to give you a better sense of their significance. For perspective and insight, Guaman Poma's original drawings and the Spanish text from his plates are presented, followed by explanation and discussion in English.

There are different pronunciations for *Waqa,* and the Spanish chroniclers recorded what they heard as best they could. Today, as with many Quechua words, there are variations of similar sounds that warrant clarification.

Waqa, is a sanctuary, a shrine, a sacred Inka or pre-Inka object of the Tawantinsuyo. Waqas were deified elements of the Hananpacha (world above), the Kaypacha (this world), and the Ukhupacha (world below), the three cosmological worlds which were worshipped by the Inkas. There were different types of waqas of different natures and with many functions; they might be deities such as with springs, rocks, trees, caves, and tóqos. Throughout Inka sacred space many waqas existed in *seq'es.* All waqas had a person called *tarpuntay* who took care of them and who communicated with them. The tarpuntay represented an ayllu or panaqa.

Fig. 7.1. A seat carved into a rock waqa at Chinchero with author Steven and field research assistant Carlos Aranibar. (Watercolor printed with permission. © 2024, Jessica Gullberg. All rights reserved)

Wakha means collapse, breakdown, or detachment, and *wakhay* means to break down, detach, or remove a part from a whole.

Wak'a is a crack or an opening in rocks that was worshipped; it is also a cleft or any opening in nature. It can as well be a cleft in the mouth, like a cleft lip.

Garcilaso de la Vega (1961 [1609]) said that waqas were "… an admirable thing, worthy of admiration for being beautiful, as it also means an abominable thing for being ugly…."

It can be said that every phenomenon that was abnormal would also become a waqa, for example Siamese twin babies, babies that were born feet first, fountains, or Apus. All three worlds, Hananpacha, Kaypacha, and Ukhupacha had their waqas. Of course, things that were ugly or abominable for Garcilaso could have been beautiful for the people of the Tawantinsuyo.

In the chronicles of Guaman Poma (2009 [1615]) there are drawings of waqas (see Fig. 7.2).

Fig. 7.2 Drawing 263 in *Nueva corónica y buen gobierno*. (Reprinted with permission from (Guaman Poma, 2010). © 2010, University Texas Press. All rights reserved)

CAPÍTVLO DE LOS ÍDOLOS, VACA BILLCA INCAP [divinidades del Inka]

/ Uana Cauri uaca / Tupa Ynga / "Uaca bilcacona! Pim camcunamanta 'ama parachun, cazachun, runtochun' ninqui? Rimari. Chaylla." ["¡Waqas, willkas! ¿Quién de ustedes ha dicho 'No llueva, que no hiele, que no granice'? ¡Hablen! Esto es todo."] / "Manam nocacunaca, Ynca." ["No fuimos nosotros, Inka."] / Con todas las uacas [divinidad tutelar local] habla el Ynga. /

/ waqa willka inkap / Waqa willkakuna. Pim qamkunamanta 'ama parachun, qasachun, runtuchun' ninki? Rimariy. Chaylla. / Manam ñuqakunaqa, Inka. / waqa /

It is important to understand that the the Inkas considered waqas to be alive and that they spoke with their waqas. Thus, every admirable thing in their lives or in their worldview was a waqa. They asked waqas to do well for them or to do them favors and to protect them from evil, danger, or disease. The Inkas believed waqas could talk, such as with the sounds that glaciers make when they are melting, or the sounds that the t'oqos make with the wind. Another form of communication was through what the waqas gave them, for example the springs and qochas on the slopes of snow-capped mountains gave them water. The Sun give them heat, food, and life; the Earth gave them water and food.

Guaman Poma de Ayala also wrote (see Fig. 7.3):

262 [264]

ÍDOLOS, VACAS DEL INGA y de los demás deste rreyno que fue en tiempo del Ynga. Es como se sigue:

Lo primero, de cómo Topa Ynga Yupanqui hablaua con las uacas y piedras y demonios y sauía por suerte de ellos lo pasado y lo uenedero de ellos y de todo el mundo y de cómo auían de uenir españoles a gouernar y ací por ello el Ynga se llamó Uira Cocha Ynga [el Inka poderoso].

Pero lo demás de cosa de Dios no le enseñó a sauer, aunque dizen que decían que abía otro señor muy grande más que ellos. Eran diablos y ací decían zupay [espíritu malo], que por tal le conocían por supay, y ancí de ellos sauían todo lo que pasaua en Chile, en Quito. De preguntar a estos supayconas [espíritus malos] tenía oficio los hicheseros pontífises llamados cunti uiza, ualla uiza.

Y ací hablaua con ellos Topa Ynga Yupanqui y quiso hazer otro tanto Guayna Capac Ynga. Y no quicieron hablar ni rresponder en cosa alguna. Y mandó matar y consumir a todas las uacas [divinidades de nivel local] menores; saluáronse los mayores. Dizen que Paria Caca rrespondió que ya no abía lugar de hablar ni gouernar[1] porque los hombres que llaman Uira Cocha [los poderosos] abían de gouernar y traer un señor muy grande en su tiempo o después cin falta. Esto le rrespondió las dichas uacas ýdolos al Ynga Guayna Capac Ynga; de ello fue muy triste a Tomi.

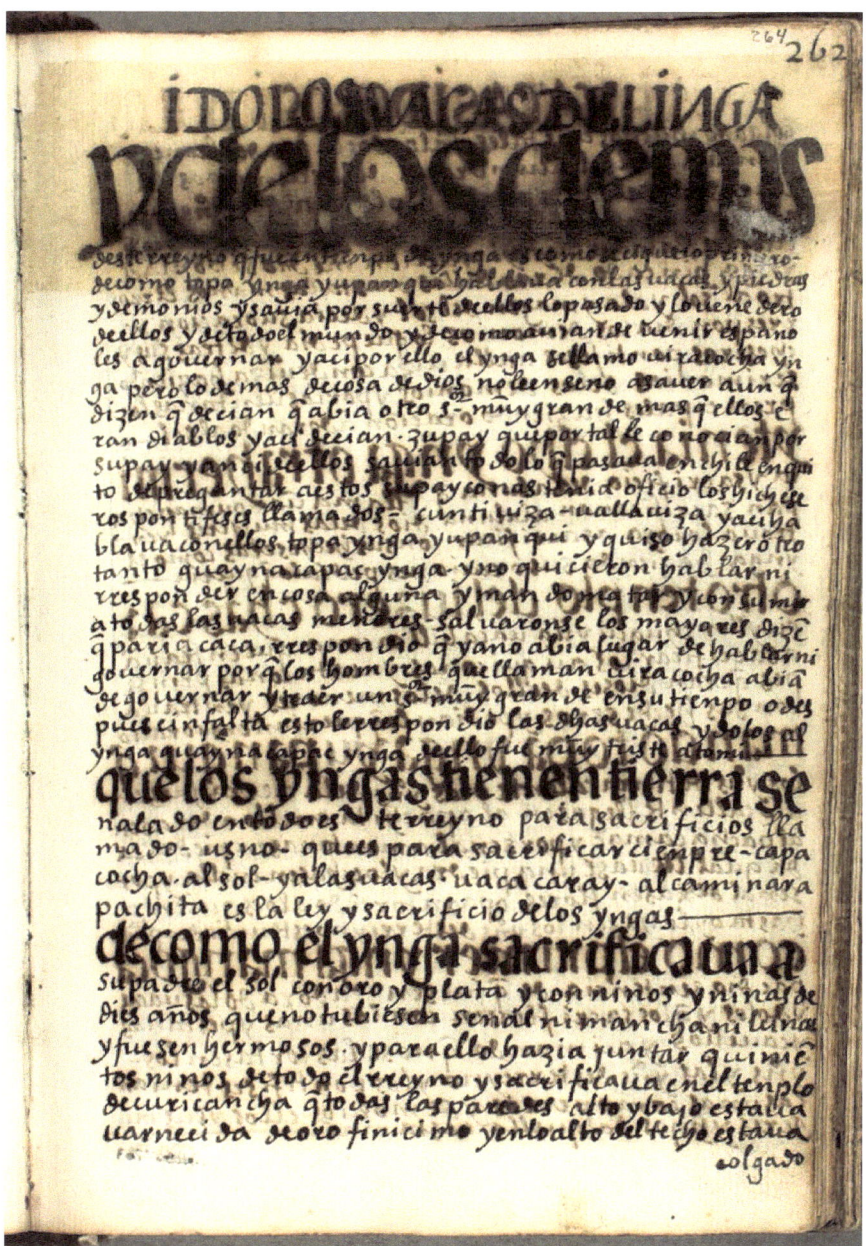

Fig. 7.3 Text in *Nueva corónica y buen gobierno.* (Reprinted from (Aranibar, 1613). The Royal Danish Academy. Public domain)

Que. los Yngas tienen tierra señalado en todo este rreyno para sacrificios llamado usno [construcción ceremonial], que es para sacrificar cienpre capac ocha [afrenta al Inka, sacrificios humanos] al sol y a las uacas, uaca caray [dar de comer a la waqa], al caminar apachita [adoratorio]. Es la ley y sacrificio de los Yngas.

De cómo el Ynga sacrificaua a su padre el sol con oro y plata y con niños y niñas de dies años que no tubiesen señal ni mancha ni lunar y fuesen hermosos. Y para ello hazía juntar quinientos niños de todo el rreyno y sacrificaua en el tenplo de Curi Cancha, que todas las paredes alto y bajo estaua uarnecida de oro finícimo y en lo alto del techo estaua

/ Wira Qucha / supay / supaykuna / kunti wisa / walla wisa / waqa / usnu / qhapaq hucha / waqa qaray / apachita /

Here Guaman Poma relates that Topa Inka spoke with the waqas, that some waqas were stones, and that the Spanish thought waqas were demons or devils and called them *Supay*. Waqas were Inka oracles, they told the past and the true; they even foretold that the Spanish would come. Waqas called supayqonas told them what was happening in other regions like Chile or Quito and those who asked or communicated with them were called Qontiwiza, Kutiwiza, and Wallawiza.

Guaman Poma also said that ushnus existed, and there were those who sacrificed themselves there for the Qapaqqocha, the Sun, or the waqas. Waqaqaray relates to walking to the apachetas. Qaray means to serve, offer food on plates or dishes to people, or give food to animals. It is also commonly written as Karay and this explains why pebbles are placed on the tops of the hills as this feeds the waqas or the apus. It is a law of sacrifice of the Inkas and therefore the Waqaqaray (feeding the Waqas) is a principle of life.

From Guaman Poma's page it can also be seen that the Qorikancha was lined with gold and silver on its walls for Father Sun. Pure children with no blemishes were sacrificed, and later what was sacrificed were the best auquenidos (llamas, alpacas, vicuñas, and guanacos).

Guaman Poma writes (see Fig. 7.4):

263 [265]

VACAS, ÍDOLOS

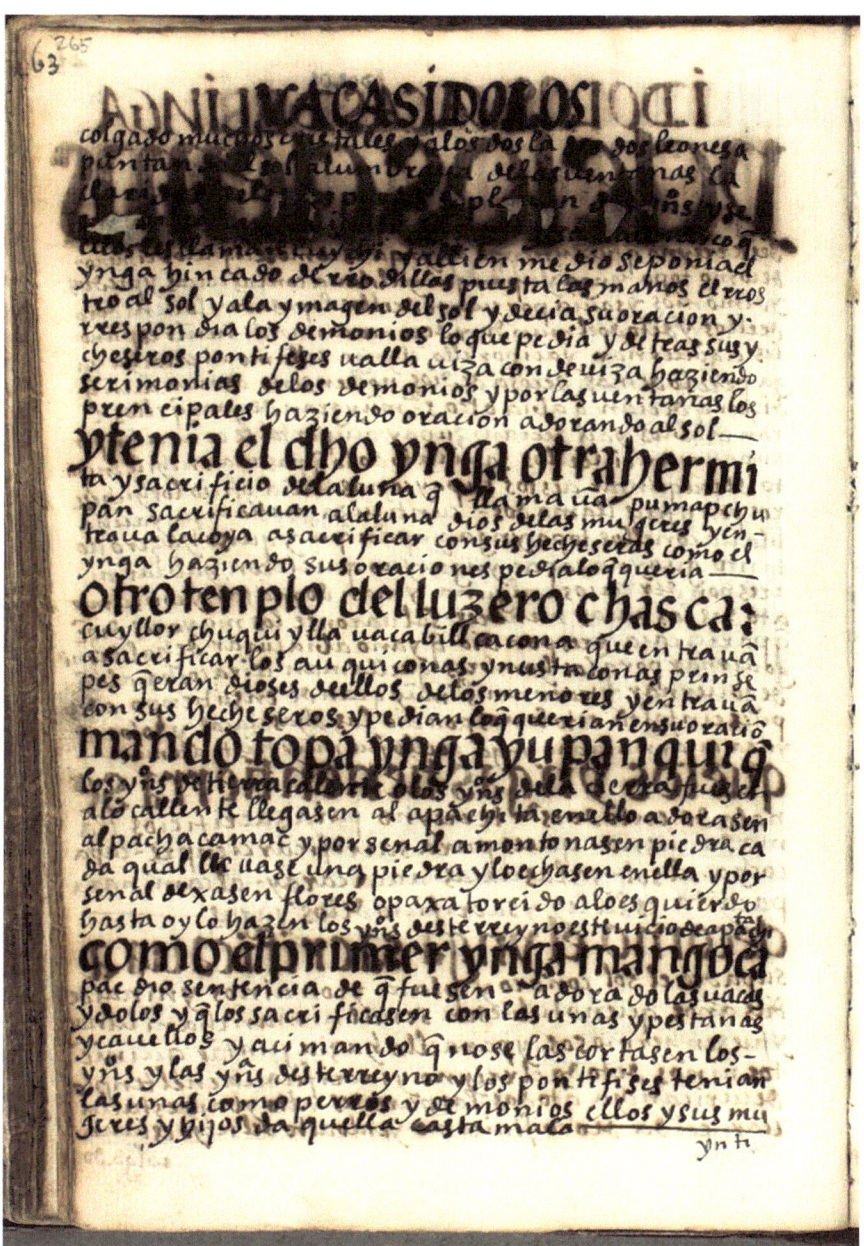

Fig. 7.4 Text in *Nueva corónica y buen gobierno*. (Reprinted from (Aranibar, 1613). The Royal Danish Academy. Public domain)

colgado muchos cristales y a los dos lados dos leones apuntando el sol. Alumbraua de las uentanas la claridad de los dos partes, s[o]plauan dos yndios y se c[…][1]. [E]ntrauan el uiento del soplo y salía un arco que ellos les llaman cuychi [arco iris].

Y allí en medio se ponía el Ynga, hincado de rrodillas, puesta las manos, el rrostro al sol y a la ymagen del sol y decía su oración. Y rrespondía los demonios lo que pedía y detrás sus ycheseros pontífeses ualla uiza, conde uiza, haziendo serimonias de los demonios. Y por las uentanas los prencipales haziendo oración adorando al sol.

Y tenía el dicho Ynga otra hermita y sacrificio de la luna que llamaua Pumap Chupan. Sacrificauan a la luna, dios de las mugeres, y entraua la coya [reina] a sacrificar con sus hecheseras; como el Ynga haziendo sus oraciones, pedía lo que quería.

Otro tenplo del luzero Chasca Cuyllor [Venus], Chuqui Ylla [a], uaca billcacona [divinidades locales]. Que. entrauan a sacrificar los auquiconas y nustacona [princesas], prínsepes, que eran dioses de ellos de los menores. Y entrauan con sus hecheseros y pedían lo que querían en su oración.

Mandó Topa Ynga Yupanqui que los yndios de tierra calliente o los yndios de la cierra fuesen a lo calliente, llegasen al apachita [adoratorio]. En ello adorasen al Pacha Camac [creador del universo] y por señal amontonasen piedra; cada qual lleuase una piedra y lo echasen en ella y por señal dexasen flores o paxa torcido a lo esquierdo. Hasta oy lo hazen los yndios deste rreyno este uicio de apachita [2].

Cómo el primer Ynga Mango Capac dio sentencia de que fuesen adorado las uacas ýdolos y que los sacrificasen con las uñas y pestañas y cauellos. Y ací mandó que no se las cortasen los yndios y las yndias deste rreyno y los pontífises tenían las uñas como perros y demonios, ellos y sus mugeres y hijos daquella casta mala.

/k'uychi / walla wisa / kunti wisa / quya / Ch'aska Quyllur / Chuqi Illa / waqa willkakuna / awkikuna / ñust'akuna / apachita / pacha kamaq /

[a] ¿Marte?
[1] Falta una palabra por el deterioro del manuscrito original.
[2] Las apachita se encuentran en uso hasta hoy en muchas partes de los Andes.

Here Guaman Poma continues about the Qorikancha, how the walls were covered with fine gold and that many crystals were hung on the ceilings. He said that two lions were depicted, but lions did not exist in South America. These instead were two pumas made of gold that were looking at the Sun.

There were two windows that allowed illumination. Two Indians blew from there to communicate, the wind came in and a rainbow came out, which in Quechua is called k'uychi. In the middle of the rainbow, the Inka would raise his hands and face the Sun. Behind the Inka were his sorcerers, Qontiwiza, Kutiwiza, and Wallawiza, and there he enunciated his prayer. The Chroniclers said that a ceremony was being performed for the demons, and that they gave them what they asked for. Prayers to the Sun were also made through the windows.

Next to this was a hermitage—a small chapel-like building with an altar that was located in an uninhabited area with a room for those who took care of its lighting and cleanliness. This was for the Moon, the Qoya, and her sorceresses, and was called *pumaq chupan* (the tail of the puma). Therefore, this enclosure could have been in the part of the Qorikancha where the convent is now located, or simply the enclosure next to the enclosure for the Sun, which is presently the Temple of Santo Domingo. From this enclosure the Qoya (the Inka's wife) asked for everything she wanted. Another enclosure in the Qorikancha was for the star Chaska Qoyllor (the planet Venus). There was a spear called ChukiIlla placed as an idol, apparently representing Mars or Jupiter, and for the Waqa Willqaqona (the sacred that gave abundance). There the Awquiqonas and Ñustaqonas, who were the princesses in the Tawantinsuyo, made sacrifices. Analyzing the word ChuiIlla, which is the composition of two words, the first Chuki, which means spear, was a weapon of war used in the Inkanato. The second part of the word is Illa, which can also mean several things. One of them is another name given to the god Wiraqocha. Illa teqsi means foundation of light and Illa teqsi Wiraqocha means supreme divinity of light.

Associated temples were erected, for example in what is now San Pedro in Qanchis. Illa is also the name of the inventor of the Khypu during the time of Inka Mayta Qhapaq. Here Illa can be a ray of light, for example one that enters through a crack or hole. Illa are also articles of precious metals, such as idols or minerals affected by lightning strikes, that had sacred virtues attributed to them. Illa can also represent an incomparable or inimitable specimen or thing. Illa could be thunder or lightning, and that is why there were the Illakancha—enclosures or temples dedicated to them. With all this, Chuki-illa could mean many things, but what Guaman Poma wrote here makes this a little clearer, because he described Ch'aska Qoyllor as Ch'aska—Lucero, bright star, a star of great magnitude; bright stars are such as Sirius with apparent magnitude of -1.47, Alpha Centauri with magnitude -0.01, and Beta Centauri with magnitude +0.6. The latter two serve as the eyes of the Inka constellation of Yaqana (the llama). Qoyllu means resplendent, luminosity, light; Ch'aska Qoyllor is the planet, Venus. They did not use a name like this for the planet Mars because Mars is red in color, precisely the color of blood caused by spears and thus the Inkas instead called Mars Chuki Illa.

It has been said that Tupaq Inka Yupanqui ordered the Runas (the people of the Tawantinsuyo) to go into the hot lands in the Antisuyo, which is largely jungle near the Amazon River. On the way they would reach the apachetas with their stones and left flowers and crooked straw on the left side of the apacheta, offered for Pachaqamaq. Pachaqamaq was the one who governed space-time, the one who created space-time, the one who orders space-time, or the one who fills space-time.

Guaman Poma wrote that the first Inka was called Manqo Qapaq. He began the worship of waqas and the Inkas sacrificed or gave waqas their nails, eyelashes, and hair; this something author Milton has always done. Every time he cut his nails he left them in an important place, either next to a tree, or in a place where they could travel—they would be like offerings to the waqas.

Guaman Poma de Ayala draws and writes (Fig. 7.5):

264 [266]

ÍDOLOS DE LOS [1] INGAS

INTI, VANA CAVRI, TANBO TOCO

/ Uana Cauri / Tanbo Toco [agujeros del tampu] / Pacari Tanbo / en el Cuzco /

[1] El encabezamiento original cruza dos páginas adyacentes, creando la frase "VACAS, ÍDOLOS DE LOS INGAS". El agregado "ÍDOLOS" en p. 266 debe haberle ocurrido pensando en la lectura de una sola página a la vez.

Inti, Wana qawri, Tamput'oqo
Wanaqawri
Tamput'oqo
Paqareqtambo

Here are three waqas, the Apu Wanaqawri, and the waqas Tamputóqo and Paqareqtambo. There are also the Waqas of the Origins, for the origin legend regarding Manqo Qapac and Mama Oqllo and also the legend for the origin of the four Ayar brothers.

Something else interesting on this page is that the Sun, the Moon, and the stars are represented, which estabishes that they are waqas. They were drawn with faces; the Sun, the Moon, and the stars were waqas and they lived for the people in the Tawantinsuyo.

Also note that both the man and the woman, the Inka and the Qoya, are asking or praying to the waqas. Additionally, there is a third character, which would have to be Qontiwiza, Kutiwiza, or Wallawiza, and also present are the Awquiqonas and Ñustaqonas.

Guaman Poma wrote (Fig. 7.6):

7 The Waqas of the Inkas 107

Fig. 7.5 Drawing 266 in *Nueva corónica y buen gobierno*. (Reprinted with permission from (Guaman Poma, 2010). © 2010, University Texas Press. All rights reserved)

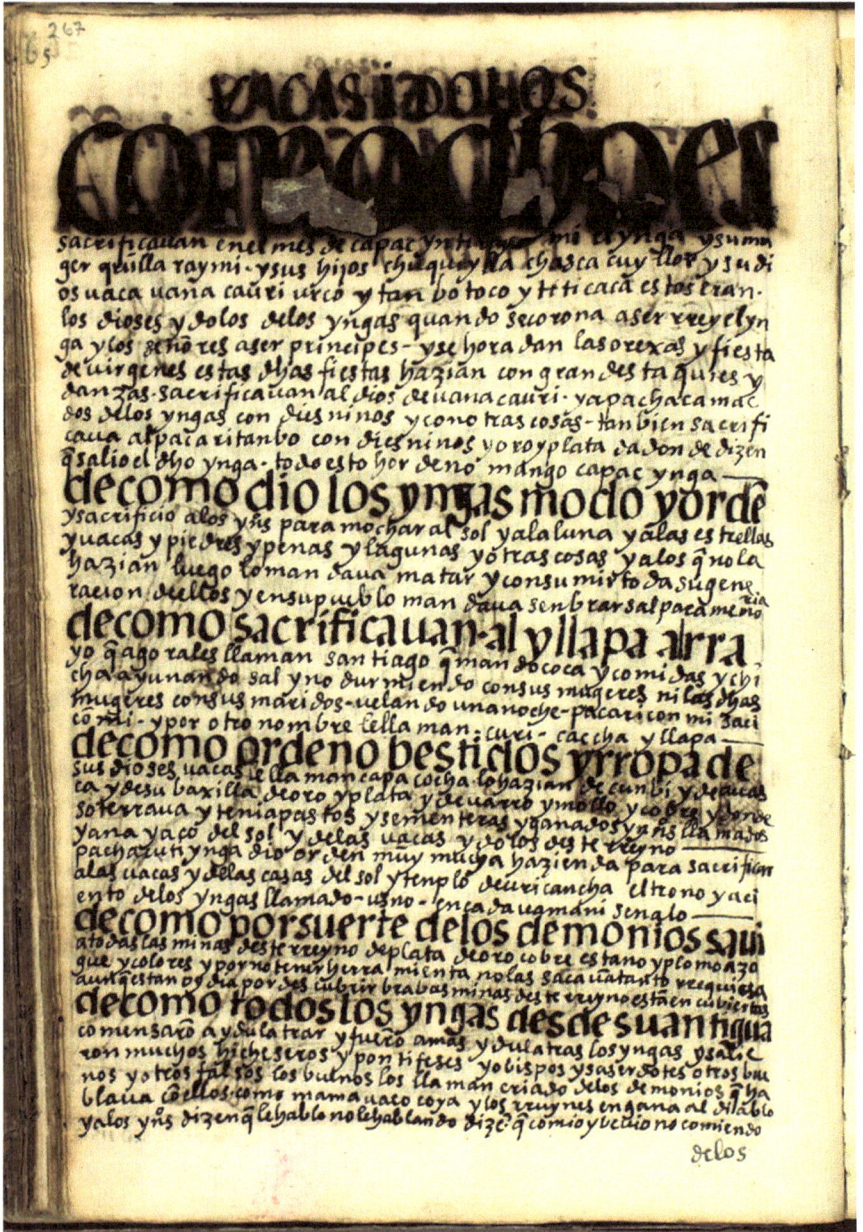

Fig. 7.6 Text in *Nueva corónica y buen gobierno*. (Reprinted from (Aranibar, 1613). The Royal Danish Academy. Public domain)

265 [267]

VACAS, ÍDOLOS

Como dicho es, sacrificauan en el mes de Capac Ynti Raymi [festejo del señor sol] el Ynga y su muger, Quilla Raymi [fiesta de la luna], y sus hijos, Chuqui Ylla [Marte?], Chasca Cuyllor [Venus], y su dios, uaca, Uana Cauri Urco [un cerro] y Tanbo Toco [los agujeros del tampu] y Titi Caca. Éstos eran los dioses ýdolos de los Yngas quando se corona a ser rrey el Ynga y los señores a ser príncipes y se horadan las orexas y fiesta de uírgenes.

Estas dichas fiestas hazían con grandes taquies [danza ceremonial] y danzas; sacrificauan al dios de Uana Cauri y a Pacha Camac [creador del universo] dos de los Yngas con dies niños y con otras cosas. Tanbién sacrificaua al Pacari Tanbo con dies niños y oro y plata da donde dizen que salió el dicho Ynga. Todo esto hordenó Mango Capac Ynga.

De cómo dio los Yngas modo y orden y sacrificio a los yndios para mochar [adorar] al sol y a la luna y a las estrellas y uacas y piedras y peñas y lagunas y otras cosas. Y a los que no la hazían luego lo mandaua matar y consumir toda su generación de ellos y en su pueblo mandaua senbrar sal para memoria.

De cómo sacrificauan al yllapa, al rrayo que agora les llaman Santiago, que mandó coca y comidas y chicha, ayunando sal y no durmiendo con sus mugeres ni las dichas mugeres con sus maridos, uelando una noche, pacariconmi, saciconmi a, y por otro nombre le llaman Curi Caccha [resplandor del oro], yllapa [el rayo].

De cómo ordenó bestidos y rropa de sus dioses uacas le llaman capac ocha [afrenta al Inka, sacrificios humanos]: Lo hazían de cunbi [tejido fino] y de auasca [corriente] y de su baxilla de oro y plata y de uarro y mullo [concha] y cobre y donde soterraua y tenía pastos y sementeras y ganados y yndios llamados yana yaco del sol y de las uacas ýdolos deste rreyno. Pacha Cuti Ynga dio orden muy mucha hazienda para sacrificar a las uacas y de las casas del sol y tenplo de Curi Cancha; el trono y aciento de los Yngas llamado usno [construcción ceremonial] en cada uamani [distrito administrativo incaico] señaló.

De cómo por suerte de los demonios sauía todas las minas deste rreyno de plata, de oro, cobre, estaño y plomo, azogue y colores y por no tener herramienta no las sacauan tanto rrequiesa. Aunque están oy día por descubrir brabas minas deste rreyno, están encubiertas.

De cómo todos los Yngas desde su antigua comensaron a ydulatrar y fueron a más ydúlatras los Yngas y salieron muchos hicheseros y pontífeses y obispos y saserdotes, otros buenos y otros falsos. Los buenos los llaman criado de los demonios que hablaua con ellos, cómo Mama Uaco, coya [reina], y los rruynes engaña al diablo y a los yndios. Dizen que le habló, no le hablando; dizen que comió y beuió, no comiendo.

/ Qhapaq Inti Raymi / Killa Raymi / Chuqi Illa / Ch'aska Quyllur / waqa / taki / Pacha Kamaq / much'ay / illapa / kuka / paqarikunmi, sasikunmi / quri k'aqcha / qhapaq hucha / qumpi / awaska / mullu / usnu / wamani / quya /

a "Han pasado la noche en vela, han ayunado."

This page is very important for the understanding of Inka astronomy. It it says that the Qapaq Intirraymi party was held for the Sun and the Quilla Raymi party for his wife, the Moon. Parties were also held for the Sun's children, who are Chuki Illa, Mars, and Ch'aska Qoyllor, Venus.

At those parties, in addition to singing and dancing, the Inkas also mocha, mocha meaning to give blown kisses. They mocha to the Moon, to the stars, to all the waqas such as stones and lagoons, and to other things like lightning.

The Catholics extirpated idolatries by repurposing waqas, for example, the Rayo called Illapa was changed to Santiago. In the time of Manqo Qapaq 10 children, gold, and silver were sacrificed here to both Pachaqamaq and Paqaritampu.

During such ceremonies the Inkas fasted, they did not eat salt or have relations with their wives and the waqas were given food, coca, and chicha to eat and drink. They kept vigil throughout night with the Nina, the sacred fire.

For the waqas they made qumbi and avasqa clothes, and they also made gold, silver, copper, clay, and mullo (seashell) basins. These basins were full of presents and were buried in the cultivation fields (semmentera) and in the pastures with cattle, which actually were auquenids. There as well were the Yanayaku Indians, also called the yanaqonas, who were of Antisuyo and were chosen to attend to the Panaqa. The Inka Pachakuteq ordered sacrifices for the waqas in the Temples of the Sun, which were the temples of the Qorikancha in each wamani (Inka provinces or administrative districts).

Guaman Poma said that fotunately the demons did not have the tools to remove the waqas' gold, silver, copper, tin, lead, and quicksilver (mercury).

Ancient Inka idolization led to the creation of sorcerers, pontiffs, bishops, and priests, both good and false. The good ones were servants of the demons and could talk to them, for example the Qoya Mama Waqo. And there were wicked sorcerers who deceived the devil and the Indians; who said that they talked to the waqas when in reality they did not and the said that the waqas had eaten and drunk, when in reality they had not.

Guaman Poma continues (Fig. 7.7):

266 [268]

ÍDOLOS I VACAS DE LOS CHINCHAI SVIVS [sic]

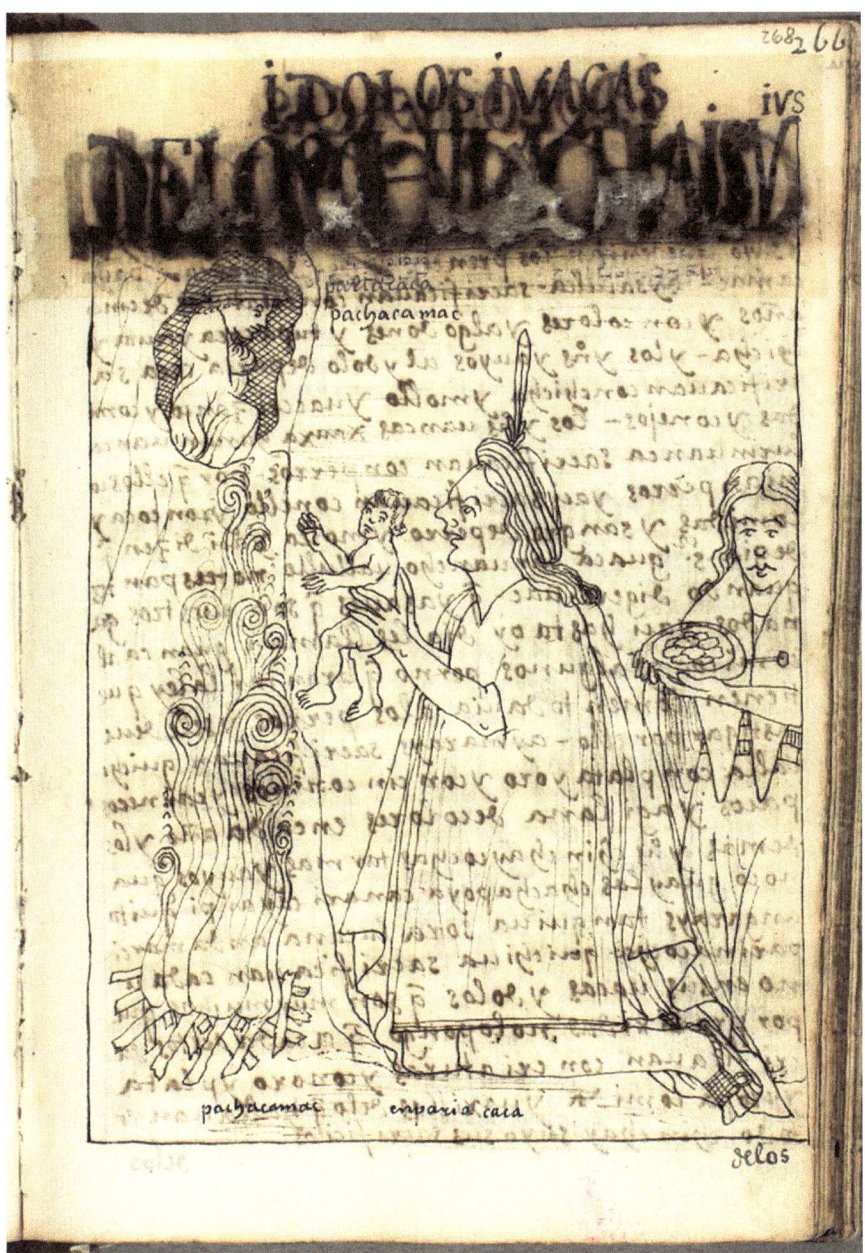

Fig. 7.7 Drawing 268 in *Nueva corónica y buen gobierno*. (Reprinted with permission from (Guaman Poma, 2010). © 2010, University Texas Press. All rights reserved)

/ Paria Caca Pacha Camac [creador del universo] / Pacha Camac / en Paria Caca /

/ waqa / Pacha Kamaq /

In this drawing there is the waqa from Pariaqaqa and Pachaqamaq in Chinchaysuyo, and it says that Pachaqamaq is in Pariaqaqa. They are offering a child and the Qoya has a vessel with other offerings. Also, in these ceremonies Nina, the sacred fire, is present.

Guaman Poma writes (Fig. 7.8):

267 [269]

ÍDOLO[S] I VACAS de los Chinchay Suyo que tenían los prencipales del Uarco, Pacha Camac, Aysa Uilca:

Sacrificauan con criaturas de cinco años y con colores y algodones y tupa coca y fruta y chicha. Y los yndios Yauyos al ýdolo de Paria Caca sacrificauan con chicha y mollo [concha] y uaccri zanco [pan remojado en sangre] y comidas y conejos.

Los yndios Uancas, Xauxa, Hanan Uanca, Lurin Uanca sacrificauan con perros porque ellos comían perros y ací sacrificauan con ello y con coca y comidas y sangre de perro y mollo. Y ací dizen que dezía: "Señor guaca Caruancho Uallullo, no te espantes quando digere 'uac' [ladrado] que ya saues que son nuestros ganados." Y ací hasta oy día les llaman Guanca, alco micoc [Wanka, come-perros]. Y algunos por no quebrantar la ley que tienen comen todauía a los perros y se le deue castigar por ello.

Aymarays sacrificauan Quichi Calla con plata y oro y con cinco niños y carneros pacos [alpaca] y agí, lana de colores en cada año.

Y los demás yndios Chinchay Cochas, Tarmas, Yauyos, Guanoco, Guaylas, Chachapoya, Cañari, Cayanpi, Quito, Angarays, Tanquiua, Sora, Lucana, Andamarca, Parinacocha, Quichiua sacrificauan cada uno en sus uacas ýdolos que son muy muchos, que por prolixidad no lo pongo. Que. a cada destos sacrificauan con criaturas y con oro y plata y rropa, comida y uaxillas de lo que hallauan en todo Chinchay Suyo sus sacrificios.

/ tupa kuka / multu / waqri sankhu / waqa qarwanchu wallullu / wanka, allqu mikhuq / paqu /

Here Guaman Poma talks about the waqas of the Chinchaysuyo. Again, he talks about Pachaqamaq, but this time he adds *Aysa willqa*. Aysa means pull and is a job on a farm with several participants. Aysay is to pull or drag, with derivatives such as aysana that are susceptible to being pulled or the place to pull or drag. There is also the word Aysaypachiquy, which in agriculture means

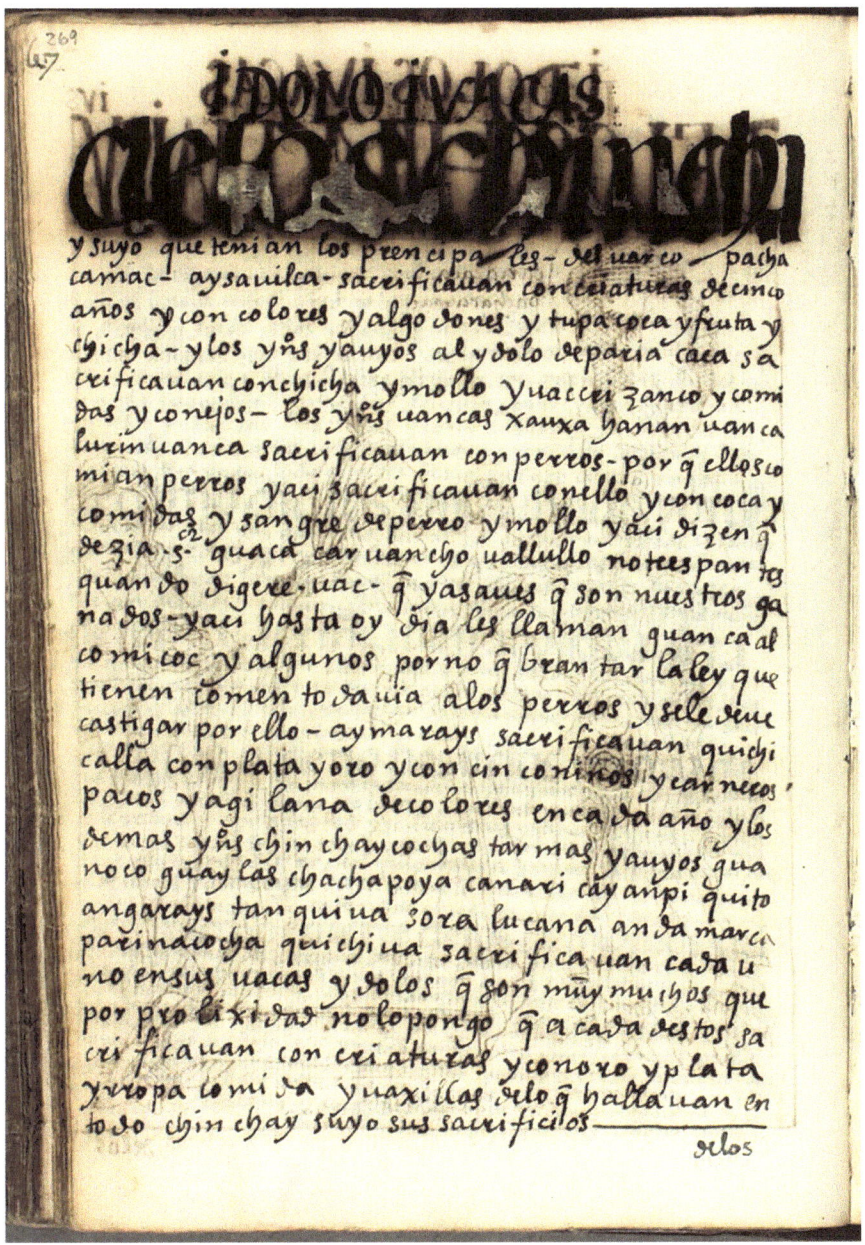

Fig. 7.8 Text in *Nueva corónica y buen gobierno*. (Reprinted from (Aranibar, 1613). The Royal Danish Academy. Public domain)

being helped by the Qollana in agricultural tasks. On the other hand, the word Willka means sacred or divine, for example Willka nina refers to the sacred fire that is used for ceremonies. Aysa Willka was a waqa for agriculture.

Other items that were given as offerings in addition to chicha were cotton, coca, fruit, rabbits, wacri zanqo, and tupa coca. Tupa is a noble object, elegant and precious with noble distinction. Tupa pachas means elegant garment and pupa coca is noble coca. Wacri Sanqo is bread soaked in blood. Sanqo, which in Qosqo is written Sankhu, is a food used ceremonies made mainly of Hak'u (wheat or corn flour that has been roasted), milk, and certain spices. The word Sankhu was used as thick or dense. For example, when Sankhu refered to a food, it was thicker and denser than liquid.

Guaman Poma also talks about the Jauja Indians, the Hanan Wanqas, and the Lurin Wanqas, but this should be Urin instead of Lurin. He said that in these Llaqta dogs were sacrificed, they ate them, and they offered them to the waqas, including offering dog's blood in molls that are seashells. This indicated that these Indians still ate dog but without breaking a law; not doing so wasn't law, it was a custom, but it was suggested to punish those who ate dogs anyway. Guaman Poma also said that the Indians had a waqa named Wanka alco miqoq, which means Wanka eats dogs. In Qosqo there is a waqa called Wanka; there is where the Spanish built a church with pilgrimages made to the Lord of Huanca.

Guaman Poma continued that throughout Chinchaysuyo in Llaqta like Tarmas, Yauyos, Guanoco, Huaylas, Chachapoyas, Cañari, Cayampi, Quito, Angaraes, Tanquiva, Sora, Lucana, Andamarca, Parinacocha, and Quichiua the Inkas offered gold, silver, precious metals, and colored wool. They fed sanqo and chili and drank chicha and blood. They dressed waqas in clothes (colored wool) and sacrifices were made to them with llamas, dogs, or rams; Guaman Poma thought llamas were rams.

In the next figure Guaman Poma writes (Fig. 7.9):

268 [270]

ÍDOLOS I VACAS DE LOS ANDI SVIOS

/ Saua Ciray / Pitu Ciray / otorongo [jaguar] / en la montaña del Anti Suyo /

/ waqa / uturunqu /

Guaman Poma tells and draws about the waqas of Antisuyo; there is Sawasiray Pitusiray, and he introduces the Otorongo or Ututunqu, a feline. The drawing can be interpreted as Sawasiray being a hill and pitusiray being two hills.

7 The Waqas of the Inkas

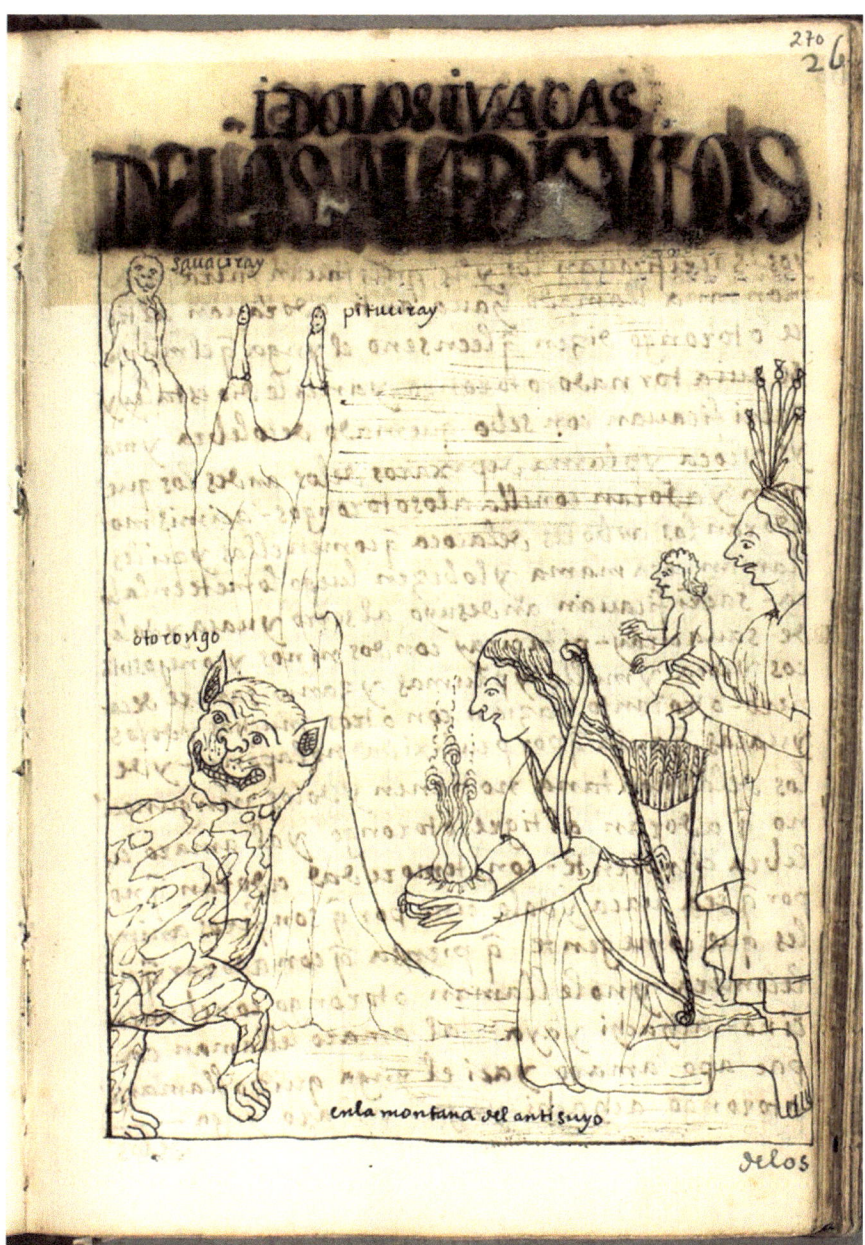

Fig. 7.9 Drawing 270 in *Nueva corónica y buen gobierno*. (Reprinted with permission from (Guaman Poma, 2010). © 2010, University Texas Press. All rights reserved)

Sawa means marriage or wedding and Siray means sewing; sewing is used to make clothing. Sira sira means sewn and overstitched on all sides; Siraq is who sews, tailor, or seamstress; Sirawa designates a rough and light seamstress or a light seam; siray kamayoq would be the tailor, the one who sews as a trade and mastery.

Sawasiray is also an aboriginal ethnic group, one of the nine ethnic groups that inhabited Qosqo upon the arrival of Manqo Qhapaq. The Sawasiray livein the location where the Qorikancha is now located.

Pitu is a couple or pair, twins equal to each other, toasted cereal flour, or a reed aerophone musical instrument very similar to a flute. Pitusira means couple, also protective goddess of couples, boyfriends, or married, represented by a couple of rocks in the snowy Pitusiray in the city of Calca. Pitukuy is to pair up, to mate, to be in pairs, and is synonymous with Iskaychakuy. P'ituy means to yearn, desire, covet, and crave; in Ayacucho it means Munay, in Ecuador this is Muana, to want, to love.

Therefore, this can be interpreted as a marriage, a link between the hills, as a light seam, or like a seam that the hills make between the Hananpacha and the Kaypacha. Pitusiray and Sawasiray might have been thought of as two twin hills, but Pitusiray was already two twin hills because Pitusira means a couple.

This then meant that the hill that is alone is the one that sews, the one that unites, and that is why it says Sawasiray and that the two twin hills are Pitusiray, the two hills that sew and connect Sawasiray.

Guaman Poma continues (Fig. 7.10):

269 [271]

ÍDOLO[S] I VACAS de los Andi Suyos:

Sacrificauan los yndios questauan fuera de la montaña llamado Haua Anti; adorauan al ticre, otorongo [a]. Dizen que le enseñó el Ynga que él mismo se auía tornado otorongo y ancí le dio esta ley y sacrificauan con sebo quemado de colebra y mays y coca y pluma de páxaros de los Andes; los queman y adoran con ella a los otorongos.

Acimismo adoran los árboles de la coca que comen ellos y así les llaman coca mama [la coca ceremonial] y lo bezen; luego lo mete en la boca.

Sacrificauan [los] Ande Suyo al serro y uaca ýdolos de Saua Ciray, Pitu Ciray con dos niños y conejos blancos y coca y mullo [caracol] y plumas y zanco, sancre de carnero. Otro tanto hazían con otros muchos ýdolos y uacas que auía, que por prulixidad no la pongo.

Y de los de la montaña no tienen ýdolos nenguno, cino que adoran al tigre, otorongo y al amaro, culebra, cierpiente. Con temoredad adoran que no porque

7 The Waqas of the Inkas

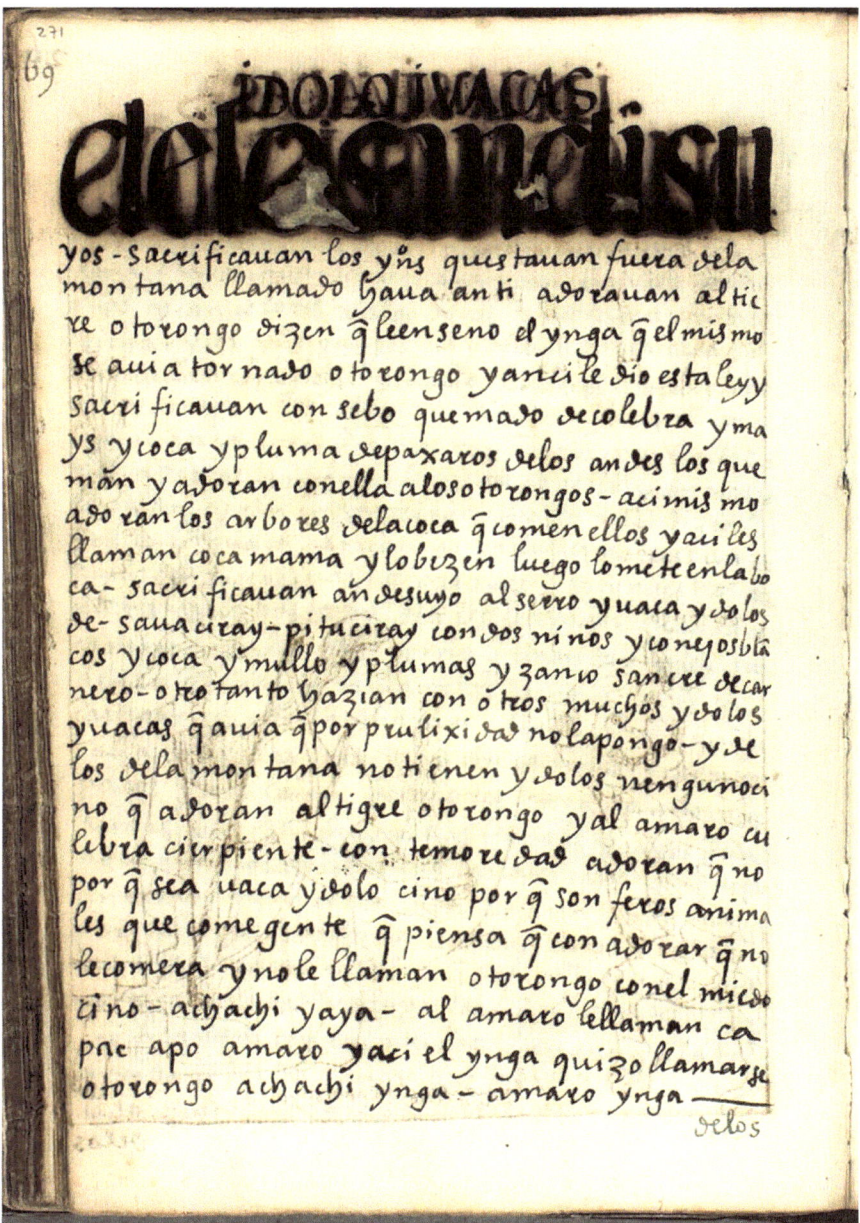

Fig. 7.10 Text in *Nueva corónica y buen gobierno*. (Reprinted from (Aranibar, 1613). The Royal Danish Academy. Public domain)

sea uaca ýdolo, cino porque son ferós animales que come gente, que piensa que con adorar que no le comerá y no le llaman otorongo con el miedo, cino achachi, yaya [abuelo, antepasado] [b]; al amaro le llaman capac apo amaro [el señor poderoso serpiente]. Y ací el Ynga quizo llamarse Otorongo Achachi Ynga, Amaro Ynga [el Inka jaguar, el Inka serpiente].

/ uturunqu / kuka mama / waqa / mullu / sankhu / amaru / qhapaq apu amaru / uturunqu achachi Inka, amaru Inka /

[a] jaguar | [b] achachi (aymara): antepasado patrilineal; yaya (quechua): antepasado patrilineal.

Here Guaman Poma talked about the waqas of Antisuyo, and that those who sacrificed were the Indians near the mountain called Hawa Anti. Hawa means a certain part, a certain place in the place, it also means after, above, on, behind, or outside on the outside. For example, Ashka Hawapi, which means in various places, haway ruphay, the Sun after the rain, or hawa runa which means foreigner. Hawa means beyond, after, either in time forward or backward, or in the place above. This then is part of the Inka worldview of the Hawapacha world, which was the time and space beyond, that is, beyond in space-time of space-time. The word Hawa Anti refers to the Indians who live beyond the Antisuyo. The Tawantinsuyo had its know limits, but this indicates that more was known and that is why Qhapaq Ñan are found as far away as Brazil.

They adored the Qoqa tree called MamaQoqa, and Qoqa was very important for the Inkas—they first kissed it before eating it. It was sacrificed with burned snake tallow, corn, coca, or bird feathers from the Andes—all of that was burned.

Sawasiray and Pitusiray, hills in Antisuyu that were waqa, had no idols. There were only waqas taken as Otorongo, the snake that they call Qapaq Apu Amaru, which means lord god or powerful snake, and thus the Inka wanted to call himself Otorongo Achachi Inka (Otorongo Inka Ancestor), Amaru Inka (Inka Amaru).

Guaman Poma draws and writes (Fig. 7.11):

270 [272]

ÍDOLOS I VACAS DE LOS COLLA SVIOS

/ Uillca Nota / carnero negro / en el Collao /

/ waqa /

7 The Waqas of the Inkas

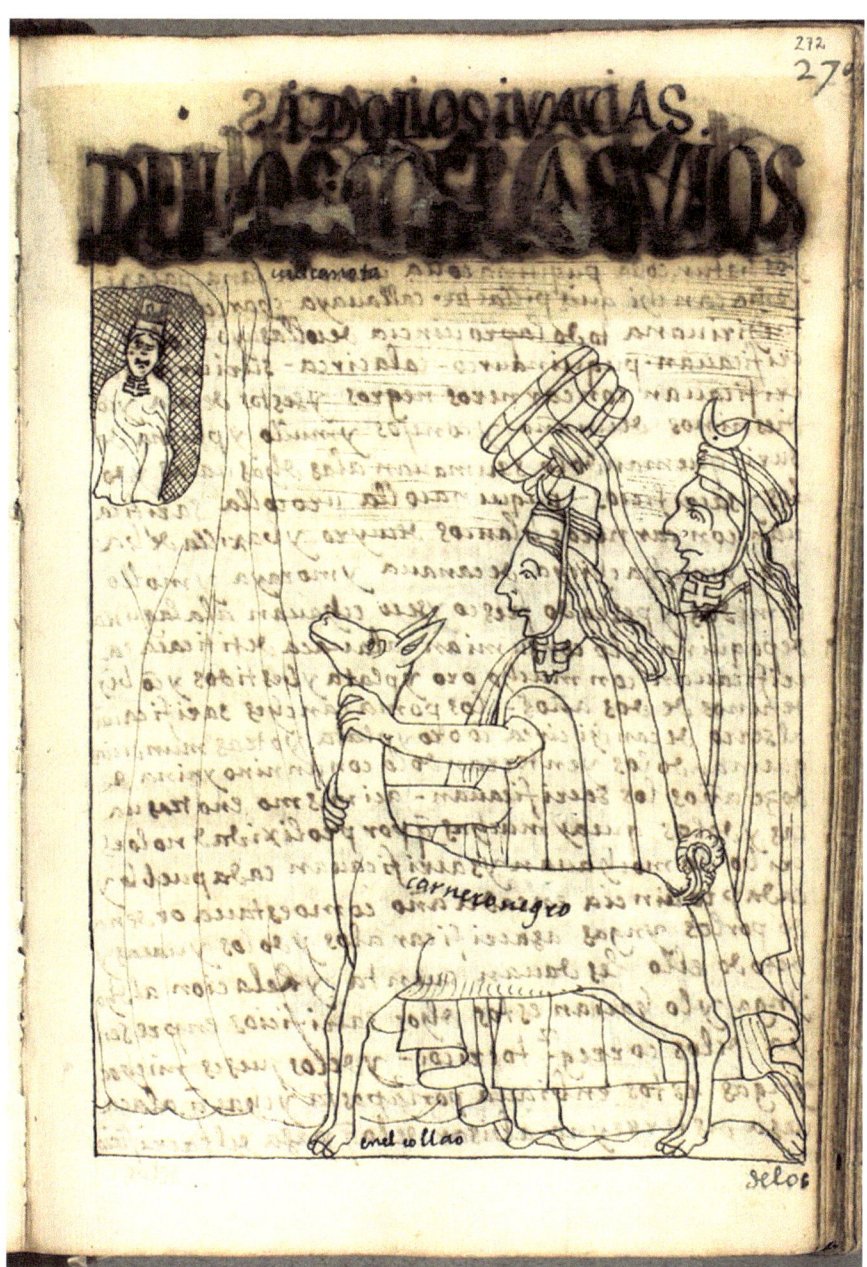

Fig. 7.11 Drawing 272 in *Nueva corónica y buen gobierno*. (Reprinted with permission from (Guaman Poma, 2010). © 2010, University Texas Press. All rights reserved)

Guaman Poma starts here by talking about the waqas of Qollasuyo and he drew the waqa Vilcanota; he drew it as a hill and this could be the hill where the Vilcanota River was born; the river in Quechua was called Willqamayu, the sacred river. He said that this waqa is in the Qollasuyo and in the drawing it can be seen that he is with a black llama that is happy. Guaman Poma again calls llamas ram. A bundle offered by the rune in the drawing can also be seen.

Next, Guaman Poma writes (Fig. 7.12):

271 [273]

ÍDOLOS I VACAS de los Colla Suyos, Hatun Colla, Puquina Colla, Uro Colla, Cana, Pacaxi, Poma Canchi, Quispi Llacta, Calla Ualla, Charca, Chui, hasta Chiriuana, todo la prouincia de Colla Suyo.

Los Idolos Waqas en el Qollasuyo son, Hatun Qolla, Puquina Qolla, Uro Qolla, Pacaqsi, Pomaqanchi, Quispillaqata que debe ser el pueblo de quispiqanchi, Qallawalla, Chgarqa, Chui, chiriwana.

Collas sacrificauan Puquina Urco, Cala Circa, Suri Urco; sacrificauan con carneros negros y sestos de coca y con dies niños de un año y conejos y mullo [concha] y pluma de suri [avestruz]; quemándolo, saumauan a las dichas uacas ýdolos y sacrificios,

Puquina Colla, Uro Colla sacrificauan con carneros blancos de cuyro [llama blanca] y baxilla de barro y mucha chicha de canaua [gramínea de altura] ^{a} y moraya [ch'uñu blanco] y mollo, comidas y pescado fresco y seco. Echauan a la laguna de Poquina y lo consumían. A la uaca de Titi Caca sacrificauan con mucho oro y plata y bestidos y con beynte niños de dos años.

Los Poma Canches sacrificauan al serro de Canchi Circa con oro y plata y otras mundicias, quemándolos y enterrándolo con un niño y niña de doze años. Los sacrificauan acimismo en otras uacas ýdolos que ay muchas que por prolixidad no lo escribo. Lo mochauan [adoraban] y sacrificauan cada pueblo y cada prouincia en cada año, como estaua ordenado por los Yngas a sacrificar a los ýdolos y uacas.

Y de todo ello les dauan cuenta y rrelación al dicho Ynga y lo hacían estos dichos sacrificios en presencia de los corregidores tocricoc ^{b} y de los jueses michoc yngas. Éstos enbiaua por la posta y chasque [postillón] a la cauesa deste rreyno abisar de lo que pasa del sacrificio.

/ waqa / kuka / mullu / suri / quyru / kañiwa / muraya / much'ay / t'uqrikuq/ michuq Inka / chaski /

^{a} cañahua; Chenopodium pallidicaule | ^{b} gobernadores reales

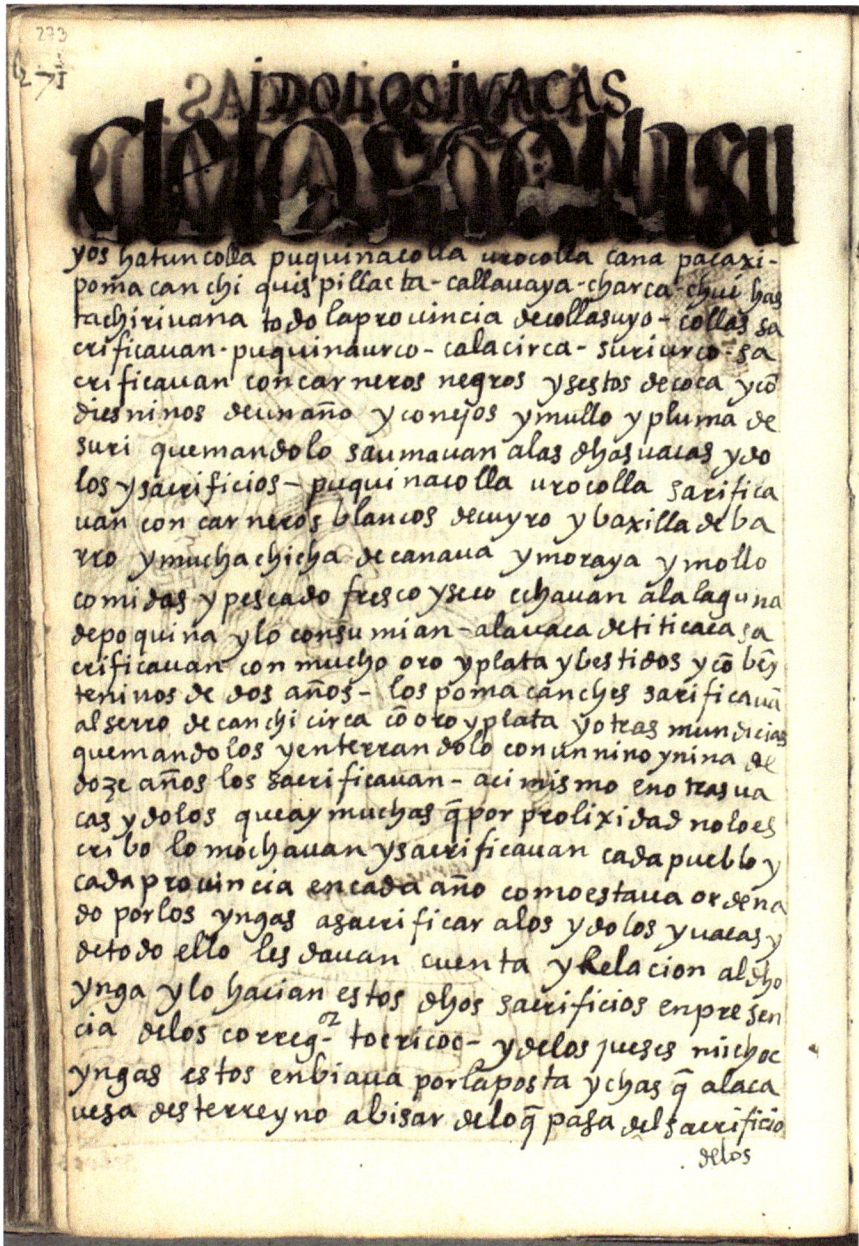

Fig. 7.12 Text in *Nueva corónica y buen gobierno*. (Reprinted from (Aranibar, 1613). The Royal Danish Academy. Public domain)

The waqas of Puquina Orqo, Qalasirqa, and Suri Orqo received sacrifices of black llamas, Qoqa baskets, rabbits, mullo, Suri's feather which is an Ostrich feather, and 10 one-year-old children. They burned all those offerings.

The waqas of Puquina Qolla and Uro Qolla received sacrifices of white llamas of Quyro, clay pottery, many chichas from Canawa and moraya, mollo, food, and fresh and dried fish. They were thrown into the Puquina lagoon and consumed.

The waqa of Titiqaqa received gold, silver, clothes, and 20 two-year-old children.

The waqa of Qachisira was offered gold and silver. Guaman Poma said that we know only the best was offered to the waqas, these offerings were burned and/or buried. Likewise, he said that a boy and a 12-year-old girl were offered to this waqa and were the Pomaqanchis. He noted that the people of Tawantinsuyo were wrong and that their wrongs justified the killings and genocide by the Spanish invaders that carried out throughout the Coordillera de los Andes; for the same reason he described the killing of children.

Guaman Poma ends by saying that the Inkas did all this in the presence of the Tuquyriqoq and the judges called michoqinkas; they recorded it in the Khypus, and the Inka was informed through the Chasky.

Guaman Poma then continues (Fig. 7.13):

272 [274]

ÍDOLOS I VACAS DE LOS CONDE SVIOS[1]

/ Coropona / en los Condes /

/ waqa /

[1] Manuscrito perforado

He talks about and drawing the waqas of Qontisuyo and he also draws the Waqadel Qoropuna, which is a volcano, and says that it is in the Qontis, which is in today's Arequipa.

On the last page dedicated to waqas, Guaman Poma wrote (Fig. 7.14):

273 [275]

ÍDOLOS I VACAS de los Conde Suyos, Ariquipa Conde, Huncullpi y Collaua Conde, Cuzco Conde, Uayna Cota, Toro, Achanbi, Poma Tanbo, Conde Suyos:

7 The Waqas of the Inkas 123

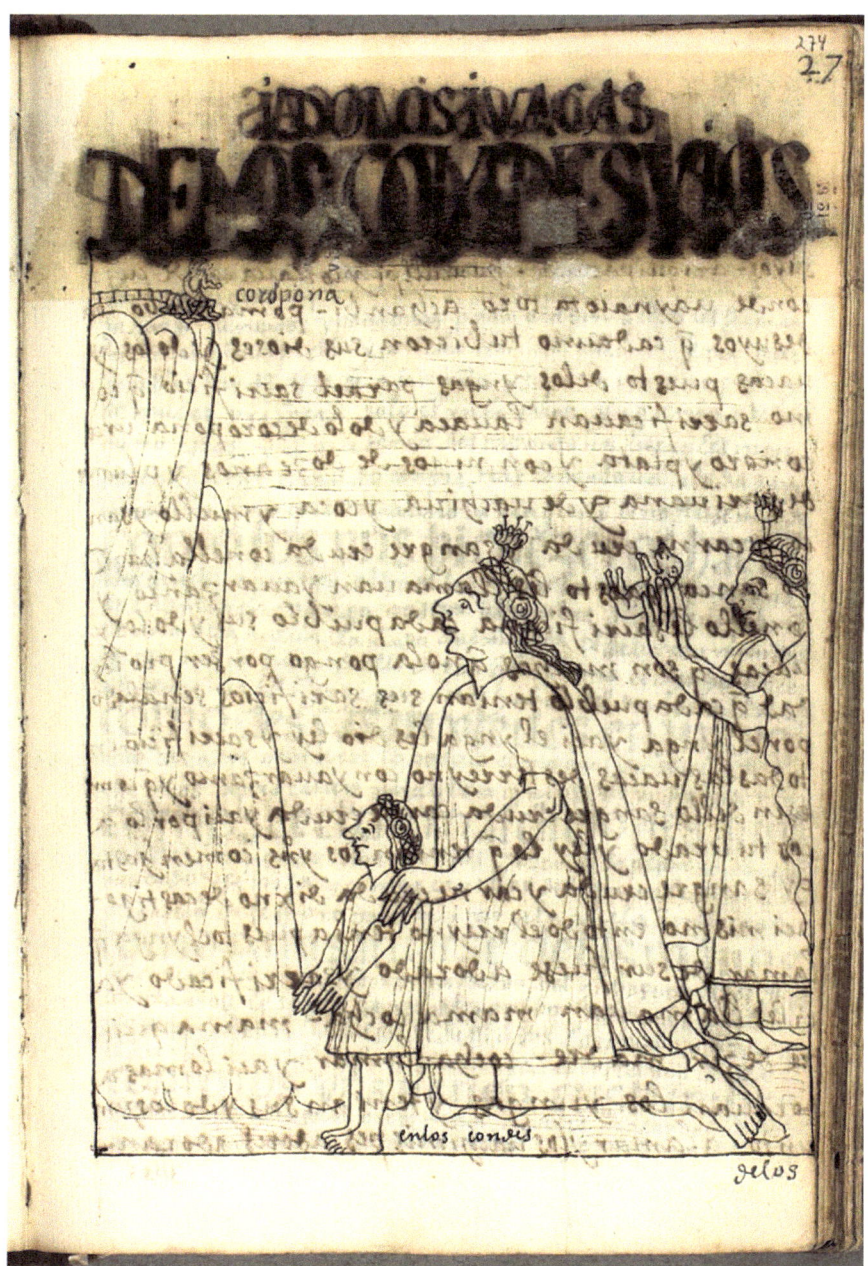

Fig. 7.13 Drawing 274 in *Nueva corónica y buen gobierno*. (Reprinted with permission from (Guaman Poma, 2010). © 2010, University Texas Press. All rights reserved)

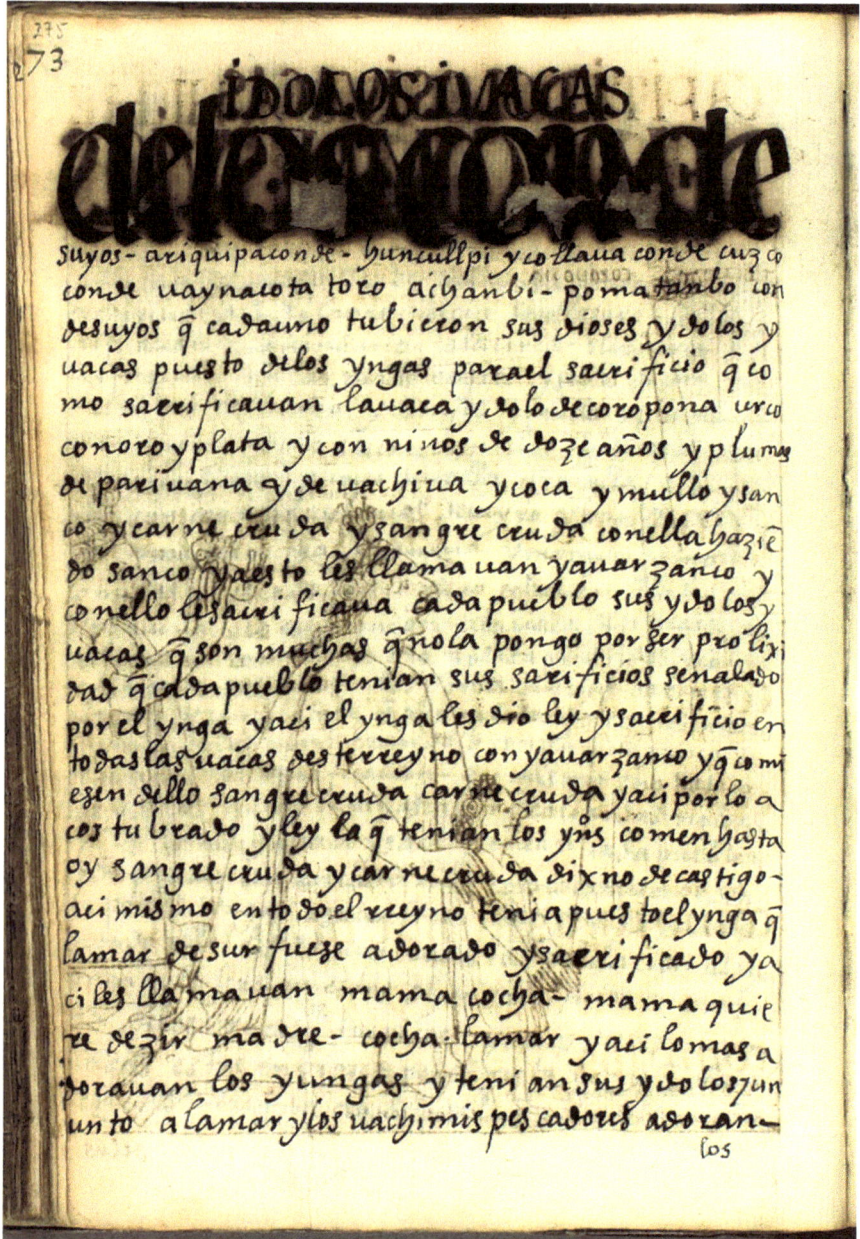

Fig. 7.14 Text in *Nueva corónica y buen gobierno*. Reprinted from (Aranibar, 1613). The Royal Danish Academy. Public domain

Que. cada uno tubieron sus dioses ýdolos y uacas puesto de los Yngas para el sacrificio; que como sacrificauan la uaca ýdolo de Coropona Urco, con oro y plata y con niños de doze años y plumas de pariuana [flamenco] y de uachiua [ganso] y coca y mullo [caracol] y sanco [sangre del carnero] y carne cruda y sangre cruda, con ella haziendo sanco. Y a esto les llamauan yauar zanco y con ello le sacrificaua cada pueblo sus ýdolos y uacas que son muchas, que no la pongo por ser prolixidad. Que. cada pueblo tenían sus sa[c]rificios señalado por el Ynga yací el Ynga les dio ley y sacrificio en todas las uacas deste rreyno con yauar zanco y que comiesen dello sangre cruda, carne cruda. Y ací por lo acostubrado y ley la que tenían los yndios comen hasta oy sangre cruda y carne cruda, dixno de castigo.

Acimismo en todo el rreyno tenía puesto el Ynga que la Mar de Sur fuese adorado y sacrificado y ací les llamauan Mama Cocha [la madre mar]; mama quiere dezir madre, cocha, la mar. Y ací lo más adorauan los Yungas y tenían sus ýdolos jununto [sic] a la mar y los uachimis, pescadores, adoran.

/ waqa / pariwana / wachiwa / kuka / mullu / sankhu / yawar sankhu / mama qucha / yunka / wachimi /

Here Guaman Poma says that in the Llaqtas of Ariqipa Qontyi, Hunqullpi, Qollawa Qonti, Qosqo Qonti (the Qonti part of Qosqo), Waynaqota, Toro, Achambi, Pomatambo, and Qontisuyos there were waqas for each placed there by the Inkas for offerings and sacrifices.

A translation for the current word Arequipa needs to consider how it is pronounced. If it is Ariqhepa, Arekipa, or Areq'epa, then Arí means yes; qhepa means after, behind or after, and following; qhepakuy is to stay, to stop voluntarily and for a more or less prolonged time; q'epa means a strident, sonorous, powerful voice; kipa means interval, flashing, discontinuous, at intervals, behind, after; and kipay means to leave interval. Therefore, Arequipa would be without making an interval—in that place one had spend time before continuing a trip.

For the waqa of the Qoropuna Urqo (Qoropuna hill) sacrifices were made with gold, silver, 12-year-old children, pariwana (flamingo) and wachiwa (goose) feathers, Qoqa, Mullo (snail), and sanqo (food based on cornmeal with milk). When sanqo was made from raw blood and raw meat, it was called yawaqsanqo since yawar is blood. And this was true for the waqas in each Llaqta of the Qontisuyo. In addition, Guaman Poma said that the Indians themselves ate Yawaqsanqo and for that reason they should be punished, thus offering another reason to the court to justify genocide. A similar justification was made when he wrote about children being sacrificed, and as a result this practice came to an end.

Guaman Poma finishes by describing a very particular waqa, the Mama Qocha (the mother Qocha) and in this case the word Qocha refers to the sea, to the south sea, and that the Yungas adored it and had their idols next to the sea and the pescadoles wamachis adored it.

He then closes with mention of the waqas in the four suyos, starting from Chinhaysuyo anticlockwise, that is, he continued with Antisuyo, then with Qollasuyo and ended with Qontisuyo.

Summary

In the Tawantinsuyo there were many waqas. There were waqas for each Ayllu and common waqas for several Ayllus or for all Ayllus. The Panaqa also had its waqas.

The waqas were asked for favors, such as doing well on trips, doing well with crops, doing well with illnesses, having enough food, and requesting that there be no natural disasters or droughts.

For the Inkas, waqas lived, and therefore waqas ate and drank and they were fed with offerings. The Inkas offered very valuable things, even children, which are the most valuable things, but not necessarily to kill them. They took the children because the waqas listened more to them because of their purity; the children had the power to make requests of the waqas, after which in Ayni those requested were granted.

Waqa were visited because they were living members of the Ayllu, and not just any members, but a very important members. Remember that for the Tawantinsuyanos the waqas lived, and they would interact with them with the principle of help and reciprocity called the Ayni, and this is the reason that so many offerings were made.

Waqas could be rocky outcrops, hills, stones, water outcrops or springs, lagoons, lakes, or the sea itself, and even the clouds; there is the Phuyupatamarka place that is a waqa for the clouds. Rainbows are also waqa. Waqas existed in solid, liquid, and gaseous states, and even a plasmatic one, as in the case of the Sun.

The waqas could be terrestrial or aerial creatures, mainly the puma (Guaman Poma described it as a lion), the otorongo or jaguar (Guaman Poma described it as a tiger), the snake, or birds such as the condor and the falcon.

Waqas could be static or mobile.

Waqas had to do with life, worldview, and with beliefs. Waqas were not only in the Kaypacha, but also in the Ukhupacha and the Hananpacha, and these worlds were waqas themselves. The waqas of Hananpacha were the the Sun, the Moon, and the stars, and also the Anka, the Waman, and the Quntur.

For each waqa there was a Kamayoq, who knew the most about it. Waqas were all admirable for their qualities or beauty and were communicated with in Ayni; they also were offered the most valuable things.

8

Inka Seq'es

Contents

Seq'es	130
The Qosqo Seq'es Emanated from the Qorikancha	133
The First Three Seq'es of Chinchaysuyo and their Waqas	136
Qollao	137
Michosamaro	137
Patallaqta	139
Pillcopuquio	139
Cirocaya	140
Sonconancay	141
Payan	142
Guaracince	142
Racramirpay	143
Intiillapa	144
Viroypacha	144
Chuquibamba	145
Macasayba	145
Guayrangallay	146
Guayllaurcaja	146
Collana	147
Nina	147
Canchapacha	148
Ticicocha	149
Summary	149

The original version of the chapter has been revised. A correction to this chapter can be found at https://doi.org/10.1007/978-3-031-67580-5_12

© The Author(s), under exclusive license to Springer Nature Switzerland AG 2024, corrected publication 2025
S. Gullberg, M. Rojas Gamarra, *Inca Cosmovision*, Astronomers' Universe, https://doi.org/10.1007/978-3-031-67580-5_8

Seq'es were also a very important part of Inka culture. Llaqta were organized with them. Polo de Ondegardo (1965 [1571]), Zuidema (1964), and Bauer (1998) described the ones surrounding Qosqo in great detail. Some are thought to include astronomical orientation and utility and Zuidema (1977) even proposed a calendar that utilized seq'es and waqas. His hypothesis is controversial. To understand Inka astronomy and culture you must first understand seq'es and their place in Inka society. This chapter offers detailed examples with the first three seq'es of Chinchaysuyo in Qosqo. Waqas were important parts of seq'es and these relationships are discussed here well.

Seq'es

According to Amauta Emilio Huaman Huillca, this word is written in singular as *seq'e*, and in plural as *seq'equna*, but it can also be written in hybrid form using the root seq' e and pluralized as with Spanish by adding an s, thus giving *seq'es*.

Seq'es emanated from the Qorikancha in Qosqo in numerous directions. The directions were for different purposes, they often were for their associated waqas, and as well were oriented for important Llaqta. Qhapaq Ñan were built to these places; therefore, some have said that certain seq'es were different directions of Qhapaq Ñan. Qhapaq Ñan left Qosqo and headed into each of the four suyus. The four main roads leaving Qosqo split into many branches. There are more than 60,000 km of Qhapaq Ñan, and the roads clearly go in all directions starting from Qosqo's center. The suyu called Qollasuyo was south and southeast of Qosqo, Antisuyo was east and northwest, Chinchaysuyo was west and northwest, and Contisuyo was south and southwest, as seen in Fig. 2.1.

Seq'es included waqas and waqas could be wanka stones, puquios, t'oqos, and paqarinas; they could also be certain trees, rivers, rocky outcrops or the Apus themselves, and certain lagoons. In other words, waqas were most places and things of great astonishment for the different ayllus of the Tawantinsuyo. Seq'es went to the different Llaqtas (regions or towns) where the ayllus lived; their waqas and paqarinas were there or relatively nearby (see Fig. 8.1).

As is common, other forms of this word have been recorded. A variant of seq'e was written as *ceqque* and was translated as raya or as aguapié de Vino or chicha, which is why the word *ceqque yascca* is found, meaning wine or chicha that has lost its vigor and has been watered or damaged. There are other variations for seq'e such as *ceqqueni* which is linear scratching or demarcation, and *ceqqueña* which is the instrument for scratching. Additional meanings say that seq'e can mean a scribble, features with amorphous lines, and it also can mean a signature. When the Spanish alphabet was first introduced with Quechua,

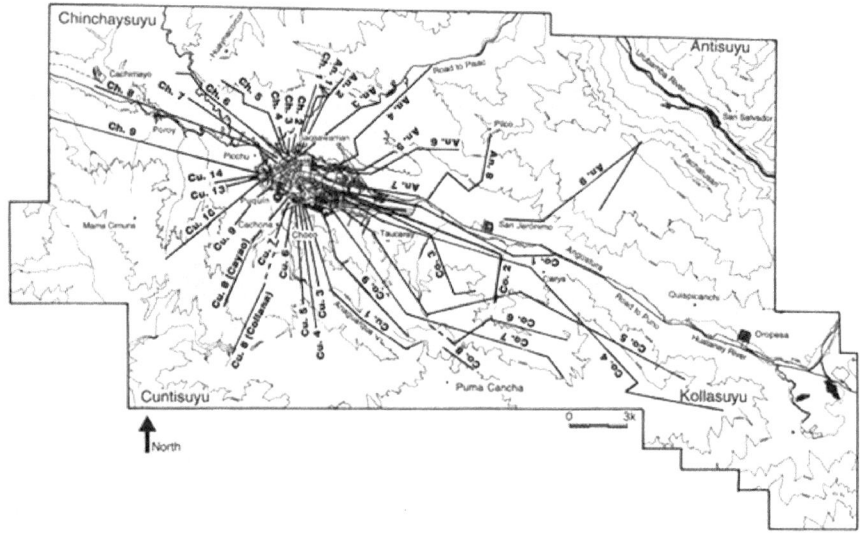

Fig. 8.1 The ceque system of Qosqo. (Reprinted with permission from (Bauer, 1998). © 1998, University Texas Press. All rights reserved)

its letters were called seq'e. Seq'e also means Jora boiled for the second time in the preparation of chicha, resulting in a very low chicha. Seq'e as well can mean an arable plot of land.

There were more than 350 waqas radiating from the Qorikancha that surrounded Qosqo. The seq'es were classified as Qollana or principal, Payan or secondary, and Kayao or origin, as were the ayllus and panaqas of Qosqo. Seq'es and waqas were distributed as follows: Chinchaysuyu had nine seq'es and 85 waqas; Antisuyu included nine seq'es and 78 waqas; Qollasuyu had nine seq'es and 85 waqas; and Qontisuyu had more with 14 seq'es and 80 waqas, making a total of 41 seq'es with 323 waqas. There were more waqas, such as four waqas without a precise location in seq'es that correlated with Chinchaysuyu. These and others are included as well in the number of waqas. In sum, there were more than 350 waqas in Qosqo, which is why this was sometimes called the Sacred Space of Qosqo. According to chroniclers such as Polo de Ondegardo, Cristóbal Molina, el Cusqueño, Bernabé Cobo, and Pedro Cieza de León, these waqas were diverse and many functions (see Fig. 8.2).

Seq'e can translate as a surface that contains scribbles or amorphous lines. And *seq'echay* meant print scribbles or amorphous lines. Likewise, *seq'echiy*, was a verb that indicated scribbling on papers or flat objects. *Seq'ena* was a punch or other instrument, such as a pencil or pen, used to scribble, scratch, or initial.

Seq'ey used in a sentence means scribble, to draw lines, and also to initial a signature. If it is desired to say that it is scribbled here or there over and over

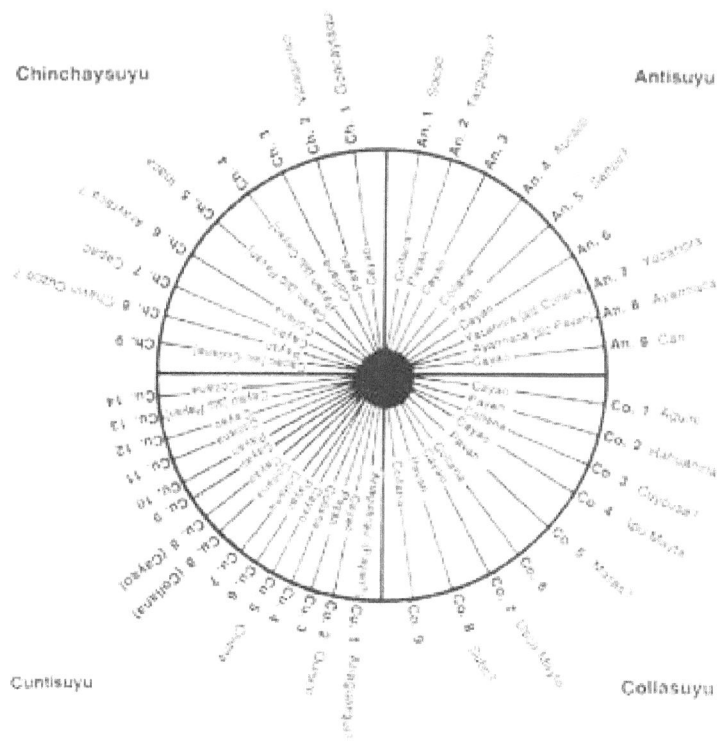

Fig. 8.2 Qosqo ceque organization. (Reprinted with permission from (Bauer, 1998). © 1998, University Texas Press. All rights reserved)

again, the word *seq'eykachay* is used; *seq'eysiy*, indicates helping to scribble on flat objects and papers or also helping to make a signature.

Of the chroniclers of seq'es, the one who said the most was Father Bernabé Cobo who wrote the chronicle *History of the Inca Empre* (Cobo, 1983 [1653]). Cobo included earlier writings of Polo de Ondegardo and it is important to note that Polo de Ondegardo gathered seq'e information by interviewing a Khypukamayoq, which means that waqas could be represented by knots in khypus. The original document that Polo wrote is lost, but its information was included by Cobo in his document.

The Qosqo Seq'es Emanated from the Qorikancha

Bernabe Cobo said a lot about the Qorikancha, first he said that many waqas from all over the Tawantinsuyo were brought there. Thus, these waqas were obviously mobile and they were idols of other ayllus in the empire "… because they are universal sanctuary of the entire kingdom, many other foreign Guacas, which were main ones of all the provinces that obeyed the Inca; which he brought to Cuzco…" (Cobo, 1983[1653]).

Cobo continued "… These gods were many, which were in the power of those of the family and awl of the King who conquered each one's province, of whom he had custody and received the sacrifices that the gods brought them … which means that the idols were in the power of the panaqas, since the King's ayllu was called panaqa; These waqas received offerings from the same ayllus that were incorporated or indexed to the Inka empire." He also said that he described only the waqas of Qosqo since while the other cities also had waqas, they imitated those of Qosco.

Waqas were located in many places: in towns, in mountains, in fields, in deserts, and in the middle of roads. During any given day of walking, people would encounter several waqas. The waqas were hills, ravines, rocks, fountains, and other things—some were temples with considerable wealth of gold and silver. Great parties and ceremonies were held at these waqas and there were guardians, Cobo called them ministers, who were the caretakers that slept in them.

According to Cobo, the most sumptuous and main temple of the Inkas was the Qorikancha. "The richest, most sumptuous, and main temple that was in this kingdom was that of the city of Cuzco, which was considered the head and metropolis of its false religion and for the most venerated sanctuary that these Indians had, and as such it was frequented by all the people of the Inca empire, who came to it on pilgrimage out of devotion. It was called Qorikancha, which means 'house of gold,' due to the incomparable wealth of this metal, which had been buried in its chapels and in the walls, ceiling and altars…." (Cobo, 1983 [1653]). This statement implies that the Qorikancha was in Qosqo and was the most important temple of the Tawantinsuyo, and that all its rooms, walls, and ceilings were lined with gold. Qorikancha translated is a patio fenced with gold since kancha is a patio, plot, or a fenced corral.

The Qorikancha was dedicated to the Sun and within it were statues of Wiraqocha, Thunder, and other idols—all the main gods of the regions of the Tawantinsuyo, in similar fashion to the pantheon of Rome.

Cobo described the architecture of the Qorikancha as "The building of this great temple was of the best factory that was found in these Indies, all inside and out of curious ashlar stones, laid with great care without mixing, and so tight, that they couldn't be any more; Although it is famous that instead of mixture there were thin silver plates placed in the joints" (Cobo, 1983 [1653]). This could be interpreted as that in the middle of the joints there were silver plates.

Juan Polo de Ondegardo described entities that were worshiped in the time of the Inkas as "Despite the Viracocha (whom they had as the supreme lord of everything and adorned with supreme honor) they worshiped also the Sun, and the stars, and the thunder, and the Earth they called Pachamama, and other different things. Among the stars, everyone commonly worshiped the one they call Collca, which we call the cabillas" (Polo de Ondegardo, 1965 [1571]; see Fig. 9.2).

Polo de Ondegardo continued "…The walls were made of strong brown stone, straight and plumb, made of very large and attractive ashlars, with some holes next to the floor like niches." Inside this fence there were many buildings; the main ones were four large structures situated in a square and were well carved; these were like chapels for Wiraqocha, for the Sun, Moon, thunder, and the other main gods. "The northern part of these enclosures, of Wiraqocha and the Sun, has been destroyed to build the temple of Santo Domingo there." Also, he said that there was an enclosure for the Mamaconas who served the temple, and more of the building for the many priests and servants who resided there (Polo de Ondegardo, 1965 [1571]).

The main room of the Qorikancha was for the Sun and it included an altar. It and the other rooms were rich with many sheets of gold. Polo de Ondegardo (1965 [1571]) said "The main room, or the main chapel and which there was the altar of the Sun and the other gods, it had incredible wealth; because instead of upholstery, it was all on the inside, ceiling and walls, dressed and lined with sheets of gold; from which we can deduce the great wealth of the temple, which was the greatest that has been found in any other in this entire New World." In addition, there were many gold idols and tableware; most was taken to Cajamarca to pay ransom for Atawallpa. Francisco Pizarro killed Atawallpa anyway.

Cobo said that the outside of the Qorikancha had gold plates that were fitted to the stone. There was only one door and when you entered it led to a small patio where the statue of the Sun was placed during the day. At night, the Inkas put it in its enclosure where it slept with the Mamaconas, who kept it company. The Mamaconas were daughters of Lords. Cobo said that they were the women of the Sun and they pretended that the Sun sat with them.

The seq'es began at the Qorikancha. Cobo wrote "Certain lines, which the Indians call Ceques, emerged from the temple of the Sun, as if from the center; and becoming four parts according to the four royal roads that left Cuzco; and in each of those Ceques there were in their order Guacas and shrines that were in Cuzco and its region, like stations of pious places, whose veneration was general to all…" (Cobo, 1983 [1653]).

From the Qorikancha the four Qhapaq Ñan left for the four suyus, and on those roads were different waqas. In Qosqo and beyond these waqas were located as stations of pious places. Cobo noted that there were many people who cared for these waqas.

"From the temple of the Sun, certain lines emerged, as if from the center, which the Indians call Ceques; and four parts were made according to the four royal roads that left Cuzco; and in each of those Ceques the Guacas and shrines that were in Cuzco and its region were in their order, as stations of pious places, whose veneration was general to all; and each Ceque was in charge of the groups and families of the said city of Cuzco, from which came the ministers and servants who took care of the Guacas of their Ceque and attended to offering the established sacrifices at their times..." (Polo de Ondegardo, 1965 [1571]).

Garcilaso de la Vega (1961 [1609]) compared Qosco with the city of Rome. He described the climate of Qosco and how its people dressed, and he continued that in the middle of the city there was a beautiful fountain (see Fig. 2.3). Garcilaso described that between the middle of the Saphi River and, crossing Qosco more or less from north to south, the first neighborhood was Qollqanpata at the foot of the Saqsaywaman hill, which could also be called Pumaorqo since perhaps instead of referring to Pumakurku Street it referred to the hill. Qollqanpata was the highest neighborhood, and that was logical because qollqas are placed in high areas; in that neighborhood there likely were several qollqas. Garcilaso continued that in the middle there was a beautiful fountain, perhaps the fountain referred to with the waqa called Canchapacha. He then described a street that passed through the middle that today is called San Agustin Street.

To the right of this neighborhood was the neighborhood of Qantupata where there were many little plants with flowers called Qantu. Then there was the Pumakunku which means the back of the Puma, along with the T'oqocahi, the hole with salt, and then Rimaqpampa which means the square where it is spoken. Next Pumaqchupan which means the Tail of the Puma and to the left looking north is called the neighborhood of Karmenqa.

de la Vega (1961) [1609]) described the neighborhoods from the northwest to the center of Qosqo—there are the neighborhoods of *Kayawqachi* and then

Chakilchaqa, and that is from where the Qhapaq Ñan leaves for the Qontisuyo. He said that near this road there were two pipes with very beautiful water and that this runs through the bottom of the Earth. It was called *Qolqemachaqway* which means silver snakes. Today *Qollqe* is gold, but Garcilaso translated it as silver because the waters are silver and flowed like snakes. Next there was another neighborhood called *Piqchu*, after this is a neighborhood called *Quillipata*, or perhaps *Quillapata*. These are elevated places, so this would be an elevated neighborhood dedicated to mother *Quilla*, who is the Moon.

Closer to the center is the neighborhood of *Karmenqa*, which is where the Qhapaq Ñan leaves for the Chinchaysuyo. Garcilaso talked about the *Wakapunku* neighborhood, and said that among many meanings, waqa means temple or sanctuary, therefore this is the door to the sanctuary. According to de la Vega (1961) [1609]) "They call it that because the stream that passes through the middle of the main square of Cuzco enters that neighborhood, and with the stream a very wide and long street descends, and both cross the entire city. A league and a half of it go where it joined the royal road to Collasuyu. They called that entrance the door of the Sanctuary or the temple, because in addition to the neighborhoods dedicated to the temple of the Sun and the house of chosen virgins, which were their main sanctuaries, they considered that entire city to be sacred and it was one of their greatest idols. From this respect they called the entrance of the stream and of the street the door of the sanctuary, and at the exit of the same stream and street they said, 'tail of the lion,' to say that his city was holy in its laws and religion and a lion in its weapons and militia."

The houses of the Inkas were located between two rivers and oriented with San Agustin Street, perpendicularly from east to west. Garcilaso described how the Spaniards distributed the houses, palaces, and sanctuaries from the slopes of Saqsaywaman to pumaqchupan, the tail of Puma.

The First Three Seq'es of Chinchaysuyo and their Waqas

Next, for greater seq'e insight detailed examples are given regarding the first three seq'es of Chinchaysuyo as decribed by Bernabe Cobo. 38 other seq'es were known to exist around Qosqo but are not discussed as this section is meant to provide insight but not to be a comprehensive compilation.

Qollao

The first seq'e of Chinchaysuyo was called *Qollao*. It included five waqas and the ayllu of *Goacaytaqui* was responsible for its care.

Goacaytaqui is also known as *Waqaytaki*, the first part of which means crying or whining. *Waqay* can mean several things, such as crowing of a rooster or an instrumental musicality, and also a cry, moan, scream, or howl. In Ayacucho it is pronounced Wajay. If combined with the word ch'uru, it then means weeper, fool, or someone who cries at every nothing. The second part of the word is *taki* which means song, hymn, or vocal music and it can be a genre of poetry whose style of sung verse had greater thematic breadth in the Inka era. There are variations such as *takichiq*, which means making people sing or teaching them to sing and this also can be used to refer to a choir director. Variations include *takichiy*, *takikuy*, and *takipanakuy*. *Takypayay* means to sing for another person in order to win their love. *Takiq* is a singer, *takiqchay* means to harmonize, or connect the chords of a melody, and *Takiy* is to sing. Waqaytaki can be interpreted in several ways, such as the crow of a rooster or someone singing in the morning with screams or moans or cries—a song of moaning or crying. If written as *Waq'atakiy* it would be singing for the waqa. This is because in that neighborhood when the rays of the Sun reach piqchu is when they would sing to the Sun in the morning and the song was like moaning or screaming or crying.

So, the ayllu, Waqaytaki, was responsible for the five waqas of the seq'e called Qollao. Bernabe Cobo (1983 [1653]) writes:

> The first was called *Michosamaro*, it was close to the foot of the *Totocache* hill, and they said they were one of those who pretended to have left with the first Inca Manco-capac from the cave of *Pacaritampu*, to whom I refer that one of the women who came out of said cave with them killed him, due to a certain disrespect he had toward her, and she turned into stone; and that her soul appeared in that same place and he ordered that they be sacrificed there; and so was the sacrifice of this very ancient Guaca; which was always made of gold, clothing, sea shells and other things, and they used to make them during good weather.

Michosamaro

Michosamaro was the first waqa of this first seq'e. The shrine was an anthropomorphic stone structure and offerings of gold, fabrics, and seashells were made to request good weather. Michosamaro can be broken down into two words, Michi and Amaru. Michi is a domestic cat, but there is no evidence that domestic cats existed in America before the arrival of the Spanish. There were wild felines, though, they could have been michi in the time of the Inkas. The word Amaru, which means snake, existed for the Inkas, and still does today. Therefore, this stone structure, like a wanqa stone, had that name because it represented both a feline and a snake.

Cobo said that this was at the foot of the *T'oqokachi* hill. *T'oqo* means cave and *kachi* means kitchen salt, esparto grass, encouragement, and it is also an honorific title for elders. So T'oqokachi would be a cave where there is kitchen salt. The T'oqokachi hill today is in the neighborhood of San Blas and the Qhapaq Ñan to Antisuyo is located there. It also is a possibility that the hill where Saqsaywaman is located was a part of the T'oqokachi hill. In front, on the opposite side from the slopes of the hill, is the path that goes to Chinchaysuyu with a waqa that was a wanqa stone with shapes of a feline and a snake, or other representations of them. Another hypothesis is that this wanqa stone was actually in the direction of the Antisuyo Qhapaq Ñan for two reasons, first because it is on the slopes of the T'oqokachi hill and second because felines and snakes are found in that direction (Cobo, 1983 [1653]).

Cobo talked about the T'oqokachi hill as if there was a cave there that he calls *Paqariq Tanpu*, and he said it is where Manqo Qapaq emerged. This might be confusing, because Paqariq Tanpu is also a province of Paruro to the southwest that is 60 km from Qosqo and is where the windows of Tanpu T'oqo are located. Perhaps Cobo confused recollections of the places he knew when he wrote his chronicle. The t'oqos were paqarinas and Manqo Qapaq came from one of them. These waqas did exist and they could have been stones or rocky outcrops (Cobo, 1983 [1653]).

Remember that *paqariq* means that it appears, originates, creates, or is born from something or someone. It is synonymous with *kamakuq*, and example can be found in the prayer *allpamanta paqariq* which translates as being born or created from the Earth.

There also is the word *paqarin* which means tomorrow, and as such it would be synonymous with *q'aya p'unchay*. Paqarin additionally means that it is created, that it appears, or that it is born.

Therefore, it can be concluded that this first waqa of the first seq'e is Michosamaro, or Michi Amaru, which was a stone wanka dedicated to, or in the shape of, a feline and snake. It was on the slopes of a hill that had a cavern.

It also is thought by some that in addition to Manqo Qapaq a woman emerged from the same cave (the principle of duality in the Inka worldview continues to be fulfilled), this woman tied to a certain conflict she had with Manqo Qapaq.

Patallaqta

Cobo (1983 [1653]) continued "The second Guaca of this Ceque was called *Patallaqta*: it was a house that Inca Yupanqui appointed for his sacrifices, and he died in it; and the Incas who later succeeded him made ordinary sacrifice here. They generally offered him all the things they consumed as a sacrifice for the health and prosperity of the Inca."

Patallaqta is made up of two words, *pata* which means above, in the upper place, or high part of a hill or house etc., and *llaqta* which means town, city, hamlet, community, region, country or territory. For example, *llaqta runa* means citizen or man of the town, and therefore *Llaqtapata*, which can also be heard as patallaqta, means town on the top.

The Patallaqta waqa is located at the beginning of the Qhapaq Ñan to Machu Piqchu. If the two waqas are considered together, this direction goes northwest but in the direction of the jungle. In that sense, this Qhapaq Ñan that begins in the *Qarmenka* neighborhood would have to continue through what is *tikatika* then go to *patallaqta* as if it were going to Machu Piqchu.

Leaving Qosqo by going to what today is called the Tikatika Arch, or the Seqa Hill, there could be a place called Patallaqta, from where there would be an overlook of Qosqo.

Since this was the house that Inka Yupanki used for his sacrifices and was the place of his death, then it must have been near the plaza of Qosqo, perhaps where Q'enqo is today (see Fig. 8.3).

Pillcopuquio

Cobo (1983 [1653]) then describes the third waqa of the first seq'e of Chinchaysuyo as "The third Guaca was called Pillcopuquio: it is a fort next to the aforementioned house, from which an irrigation ditch emerges; And the Indians say that having built that house for sacrifices, Inca-Yupanqui ordered that water to come out of there, and then decreed that an ordinary sacrifice be made to it."

Pillqopuquio is made up of two words *Pillqo*, or *Pillku*, which means different shades or also a mixture of many colors with a predominant color. For example, *yanapillku* means black tinged with other colors, and Puquio, or Pukyu, means a spring or other source of water.

This shrine was a fountain that was next to the mortuary house of the Inka Pachakuteq, where a ditch emerged. It is said that this also served as the

Fig. 8.3 Patallacta, between Qosqo and Q'enqo

seventh waqa of the eighth seq'e, *Ayarmaka*, of Antisuyu. This shrine was a fountain that was near the Llaqta of *Qoraqora*.

Cirocaya

Next is the fourth waqa, described by Cobo (1983 [1653]) as "The fourth Guaca was called Cirocaya: it is a stone cave, from which they believed hail came out; Therefore, while they were afraid of it, they all went to sacrifice in it, so that it would not come out and destroy their crops."

There is not a direct translation for *Cirocaya*, but there are variants, and this is the union of two words, for the first *Ciro*, *Siru* means a net for hunting birds. Siruka means little thief, mischievous, or searcher in corners of a house. *Chiru* means side, lateral part, elbow, cheater, poor, or bird of the eastern region. For the second word *Caya*, *kay* means to be, exist, or have and thus there can be phrases like *ñoqa kani*, I am; *qolqe kan*, there is money; *kay wasi thuniukunqa*, this house is going to collapse; *kaypi*, here; *kaymanta*, from here; *kayninta*, over here; and *kayniqman*, this way. Another word from *kay* is *kayawkachi*, which is a proper name for a neighborhood in Qosqo that is next to *Pumaqchupan* where Avenida del Sol and Tullumayo currently converge and at the confluence of the Sap'i and Tullumayu rivers. Other variations for caya are *qayara*, which is a plant that grows between 3700 to 4000 meters above sea level; *qaya*, which is an adverb synonymous with wayma, and wayna, which has to do with the past because it means *joben*, young man, teenager, and is synonymous with *q'aqo*. *Qayna* is an adverb that means last time or on a date prior to the present. For example, *qayna p'unchay*, yesterday; *qayna wata*, last year; and *qaynalla*, recently alone. Qayna here is time and kay with its suffixes kaypi or kaymanta is a place. All qay and kay serve to indicate something about the Pacha.

Q'aya indicates future, near, or coming time, for example *q'aya wata* which is coming or next year. *Qaya* has to do with past and *q'aya* with future, philosophically and physically it was thought that time is an illusion and that past, present, and future are one and intertwined with space.

Therefore, Cirocaya could be interpreted as *Chiruqaya* to the past or *Chiruq'aya* to the future. From a Western perspective it would be a waqa that is to the side, and it would separate either to the future or to the past. However, from the Quechua perspective this would be a waqa that is to the side in space-time.

Sonconancay

The fifth waqa is *Sonconancay*, and Cobo (1983 [1653]) wrote "The fifth and last Guaca of this Ceque was called Sonconanacay: it is a hill where it was very ancient to offer sacrifices for the health of the Inca."

Sonqo means heart and *Sonqonanay* is a doctor of the heart who causes compassion. *Urpi Sonqo* is heart of a dove, very affectionate. *Rawraqsonqo* means it burns a lot and *tukuysonqo* is with all my heart. *Millaysonqo* means bad character and *Sonqo Ñak'ay* is a place for the sacrifice of hearts. This shrine was a hill where people prayed for the health of the Inka. The second part of the name could come from such as *Ñaka* which is difficulty, sacrifice, or hardship; *Ñakan hamuchani* which means I am coming with difficulty; or *Ñakay* which is cussing, wishing evil on another person out of revenge, or blaspheming. Sonqoñak'ay can be translated as a place where sacrifices of hearts are made or a place where someone goes with difficulty and with all one's heart. This might possibly be what today is called *Tambomachay* (see Fig. 8.4).

Fig. 8.4 The cave on the hill at Tambomachay

Payan

The second seq'e of Chinchaysuyu was *Payan* and its first waqa was called *Guaracince*.

Guaracince

As Cobo (1983 [1653]) described "The first Guaca was called Guaracince, which was in the plaza of the temple of the Sun, called Chuquipampa; It was a small piece of plain that was there, in which they said the Earth tremor was formed. They made sacrifices in it so that it would not tremble, and they were very solemn; because when the Earth shook, children were killed, and sheep and clothing were ordinarily burned, and gold and silver were buried."

Examining further, *Chuki* means spear, hard, consistent, or strong. *Chukipanpa* includes pampa, an esplanade of spears. This is the pampa of the archaeological complex of Saqsaywaman.

8 Inka Seq'es 143

Fig. 8.5 Saqsaywaman with its zigzagged walls as viewed from the north

From Guara in Guaracince, *Wara* means scarf, mud flap, underpants, overpants, or pants worn by working men in the Inkario. From this comes the *Warachikuy*, a ceremony of virility in which young people pass to the stage of puberty or joben.

From cince in Guaracince, *sinsi* is laughter. Sinsi sipas joben is a woman who laughs a lot. Sinsiy is to laugh often and is a synonym of Cheqchiykachay.

Therefore, Guaracince can be translated in several ways:

Warasinsi is a waqa related to the warachikuy since at that celebration the new jobenes wore the wara garment.

Warasinri was a waqa that had to do with the virility of the joben. One of the tests of the Warachicuy was to stand in a column and pierce a target with an arrow.

Warasinchi because the warachikuy festival is a force of courage, of strength, of bravery. This first waqa, *Warasinse*, must have been a wanka stone, like a phallic object, and it must have been in the center of the Saqsaywaman esplanade which was called *Chukipanpa* (see Figs. 8.5).

Racramirpay

The second waqa of Payan was called *Racramirpay*. According to Cobo (1983 [1653]) "The second Guaca was called Racramirpay: this was a stone that was placed in a window that was a little below where the Sanagustin convent is now, whose history they relate in this way: that in a certain battle that Inca gave Yupanqui to his enemies, an Indian appeared to them in the air and helped him defeat them, and after achieving victory, he came to Cuzco with the said Inca, and sitting in that window, he turned into stone; which since that time they have worshiped and made an ordinary sacrifice to it; and it was particularly solemn when the Inca personally went to war, asking him to help the King as he had helped Inca-Yupanqui in that war."

Analyzing Racramirpay, raqra means a crack, split, or cracked. For example, raqra manka is a cracked pot. Raqraq is an object that suffers cracking. Raqray is a crack. Miray is reproduction of people and animals, also profit or interest. Miraywa is a fertile person or animal. Mirka is a freckle, dermatosis, or blemish on something, synonymous with mirkha. Therefore, Racramirpay can be translated as RaqraMirkay which could be a natural crack in a rock with spots perhaps of gold. That rock could have been in a window.

This waqa would likely be where the San Agustin hotel is today, on the street of the same name.

Intiillapa

The third waqa was a gold idol called *Intiillapa*. This waqa's story, according to Bernabe Cobo, is that in battle Inka Yupanki saw an Indian in the air who helped to win the battle; following the battle when they returned to Qosqo the Indian turned into stone. Perhaps the stone had the shape of a human, but most likely not. This waqa likely was in a window, like the ones in the Qorikancha. The waqa was asked for the Inka to do well in war. Since there were few conflicts, it also might have been asked for good negotiations to annex more regions for the Tawantinsuyo.

Viroypacha

The fourth waqa was called *Viroypacha*. Cobo (1983 [1653]) described "The fourth Guaca was called Viroypacha: it is a pipe of reasonable water, which was named Guaca Inca-Yupanqui; "I begged him for the quietness of the Inca."

Wiraqocha, which means foaming lagoon, was the supreme god of Tawantinsuyan mythology and *pacha* means space-time.

Once again, there can be several interpretations. *Wiruypacha* which literally would be like sucking space-time, can also be *Wirapacha*, which has to do with tallow, which is the fuel of fire, and with space-time. This can be interpreted as the fuel to generate space-time.

The exact location is uncertain, but it was in the vicinity of Saqsaywaman.

The *Wikakirao* ayllu was responsible. This was a natural water channel and came from a puquio. It could give the sensation that that water was being sucked.

This waqa was also dedicated to the Inka and was asked for his stillness or tranquility, so the Wirapacha would have something to do with stillness and tranquility.

Chuquibamba

The next waqa was called *Chuquibamba* and according to Cobo (1983 [1653]) "The fifth Guaca was a plain called Chuquibamba, which is next to the fortress; "They sacrificed him like the others."

Ch'uk means to express the silence and stillness of people and animals. Ch'ukuy is to gather, reduce length, or make folds when sewing and therefore reduce the length. Chuki is a spear or hunting weapon used in the Inkanate that is hard, consistent, tough. Chuku is a hat, headdress, or cap, or unpopulated land in the middle of the forest. Chukiy is to throw or use a spear to hunt. Bamba is a plain or pampa. Panpa is plain, a plain region.

Cobo did not say where this plain was, perhaps it was close to the plain of Saqsaywaman, but if interpreted as *Chukipampa*, then it would have been the same esplanade as Saqsaywaman.

Macasayba

The sixth waqa was called *Macasayba*. Cobo (1983 [1653]) wrote "The Guaca sect called itself Macasayba: it was a large stone that Inca-Yupanqui placed next to the plain of Chuquibamba, and ordered that veneration and sacrifices be made to it for the health of the King."

Maqa means punishment, spanking, hitting. Maqana was a weapon of war of the Inkas. Maqanakuy means contention, to fight. Maqay is to punish. Maka means tinajón, aribalo, unsociable, elusive; a tasteless thing; it is also a current breakfast based on potato flour. Maki is a hand. Mak'a is a human or ape arm. Saywa is a column, pile, boundary marker, or boundary sign; a synonym of will; a plant of erect growth.

This waqa was cared for by the ayllu *Wika Kirao*. It is a wanka stone, a *Saiwa* that the Inka Yupanqi had placed near the Saqsaywanan esplanade.

Since the waqa was asked for the health of the Inka, this seq'e could be part of asking everything for the Inka and the Tawantinsuyo.

Guayrangallay

The seventh waqa was called *Guayrangallay*, and according to Cobo (1983 [1653]) "The seventh Guaca was a quarry called Guayrangallay, which is above the fortress, in which they made sacrifices for various respects."

From Guayrangallay, wayra means wind and *cririwayra* is a cold wind. Wayran also is wind. Wayrachikuy is to vent or aerate. Wayrachinais a place where venting is done. Qoriwayrachina is a place where gold dust is sold.

This is north of the slide in Saqsaywaman, where there is a stone quarry.

From Guayrangallay, *Wayranqhallay* could mean that it is the wind that is cut, a place where the wind is cut, which would serve to protect oneself from the wind.

Wayranqalla could mean a circular wind. Wayranqallarichiy is the wind that begins. WayranKhalla is wind that is cut. WayranQ'alla is a slice of wind, it is the wind that when passing over the rocks is cut like a slice, since it passes everywhere. WayranQ'allay is a slice or cut to the wind. WayranQ'ala is the naked wind.

Guayllaurcaja

The eighth waqa was called *Guayllaurcaja* and was described by Cobo (1983 [1653]) as "The eighth and last Guaca of this Ceque was called Guayllaurcaja: it is a little pass that is made in the middle of a hill, where Viracocha Inca sat down several times to rest, climbing the said hill; and from that time and by his command it was considered a shrine."

From the first part of Guayllaurcaja, *Waylla* means land with spring and grass. Wayllapanpa is a pampa with a lot of grass; this also is a typical dance of Qosqo.

WayllaWaylla is a land with lots of grass mixed with forests. Wayllapanpa/Wayllar populations are areas where there are springs with grassland; today in the district of Sanjeronimo there is a recreational place that meets these characteristics, and it is called wayllar. Wayllani is an extensive meadow with grass.

The second part qata means blanket, qasa is frost, cold, winter, a synonym of Chiri or khutu, and qasapacha is frozen time.

Qaray is to serve, a synonym of karay. Qara means body, skin. Qala has to do with chichi that is naked, qalato refers to that which is naked. Qachi is dispersion. Qati is continuous, correlative, followed in a chain or row; and is synonymous with qatilla; in Ayacucho they call it seje or sinri. Kaka means luminous, bright, a narrow-mouth clay jar, a mother's brother. K'aka is a crack. K'akara means fleshiness, or crest, fleshiness on the head of birds such as the Quntur. Kaq means what is or what exists. Kay is to be, exist, have ukillus such as the wayllabamba district in Urubamba or urqos of the lucar near the archaeological park of Pikillaqta.

Guayllaurcaja could be *Wayllarqasa*, which would be the waylla es frio, that is, the small mountain pass nearby that has a grass with springs and is cold.

Wayllark'aka is a small door with a crack. Wayllarkaka is a shiny porch or porch with many rocks.

Guayllaurcaja was located between the Saqsaywaman hill and the T'oqocahi. The word *portezuelo*, which is no longer used today, is a diminutive of port, and port is a place where ships arrive, but it also means a place where you can cross an obstacle, like a mountain or a river. So, since all these waqas are in Saqsaywaman, this waqa would be in the middle of the Saqsaywaman complex at the top of the T'oqocachi hill where the White Cristo is today. At a small mountain pass there is a pampa where there are several puquios, several springs and the waters join to form a small river called Tullumayu that flows from Hananqosqo to Hurinqosqo and then joins the Watanay River. This is a good place to sit and contemplate, as part of both Saqsaywaman and Qosqo, and it must be close to its palace.

Collana

The third seq'e was called *Collana* and it had 10 waqas; here the first three are included as examples for insight.

Nina

According to Cobo (1983 [1653]) "The first was called Nina, which was a brazier made of a stone where the fire was lit for sacrifices, and they could not take it from anywhere else; It was next to the temple of the Sun, and great veneration was held for it and solemn sacrifices were made to it."

Nina means fire, candle or fathom, what can be said, what must be said, for example Nina and awray—the fire in flames; ninak'anchay—fire luminary; nina sut'inta—who must tell the truth.

Nina nina is a lot of fire. Ninaq is related to fire. Ninay ninay means very incandescent, increased temperature, is synonymous with rupha rupha.

Nina was attached to the north wall of the Qorikancha. It was a stone brazier that was next to the Qorikancha, where offerings like the qoqa were burned. The name Nina was used because fire was sacred to the Inkas. Fire gives heat at night, heat for cooking food, and also brilliant light. The Inkas did not know how fire was physically produced, so they saw it as magical and sacred like an apu or a waqa.

Canchapacha

The second waqa was *Canchapacha.* Cobo (1983 [1653]) describes "The second Guaca was called Canchapacha: it was a fountain that was on Diego Maldonado Street, to which they made sacrifices for certain stories that the Indians told."

Kancha means patio, a canchon surrounded by a wall, a corralon, a sports field, a livestock corral, a manger. K'ancha is light, luminosity, brilliance, splendor. K'ancha k'ancha is between lights, semi-illuminated. K'anchachi means 'that produces luminosity,' a source of light or luminosity. Nina k'anchachi is the light of the fire, the fire that is a source of light, what commands illumination. K'anchachiq means 'that illuminates,' enlightener, the one who illuminates. K'anchachiy is to illuminate. K'anchay is light or shine; also offering that must be given to the dead for three consecutive years.

Pacha means space-time. Phaqcha is a jet, trickle, a waterfall of water or other liquid that falls from a certain height. Phaqchay is a drip, a drop of a liquid.

This can be interpreted in several ways: Kanchapacha is the courtyard of space-time, it also is a fountain or paqarina, it can be like the paqarina for the origin of space-time, and water sources were paqarinas for the origin of life. Kanchaphaqcha is a courtyard with a water fountain and k'anchapacha means the resplendent light of space-time.

Canchapacha was on what is now San Agustin Street. From Saqsaywaman flows a stream called *Tullumayu* that crosses Qosqo. Between it and the Saphy River are all the palaces of the Inkas who have directed the Tawantinsuyo. The fountain would likely have been on what is now San Agustin Street. According to Garcilaso de la Vega (1961 [1609]) "At midday (south) of the Iglesia Mayor, in the middle of the street, are the main stores of the richest merchants." Behind the church were houses that belonged to Juan de Berrio, and others. Behind the main stores were houses that belonged to Diego Maldonado, who was called The Rich, because he was very wealthy due to being one of the first conquerors. In the time of the Inkas that place was called *Hatuncancha,* and it means big neighborhood. Inka Yupanqui's houses were there. The neighborhood is called Puca Marca.

Ticicocha

The third waqa of Collana was called Ticicocha. As described by Cobo (1983 [1653]) "The third Guaca was another fountain called Ticicocha, which was inside the house that belonged to the said Diego Maldonado. It was this fountain of the Qoya or Queen Mama Ocllo, in which very large and ordinary sacrifices were made, especially when they wanted to ask something from the said Mama Ocllo, who was the most revered woman among these Indians."

Teqsi means foundation or base. Teqsi Wiraqocha is a Fundamental Lord, the main god of all the Inka gods, the base god. Teqskhay is to lay the foundations. Teqsimuyu means it is the world, the celestial sphere, the planet Earth, the orb.

Qocha is a lagoon or lake. This waqa was one of the main lagoons in the Inkario, perhaps it was a lagoon that was formed from the water that came out of a very important puquio; it could even have been the one next to the Qorikancha—a puquio because it was the source of Mama Oqllo who was the wife of Manqo Qapaq. Mama Oqllo was the main queen of Tawantinsuyo since she and Manqo Qapaq left Lake Titiqaqa together. Tayta Inti gave them a gold bar and told them that where this bar sank, they would found an empire. For this reason, Mama Oqllo became the primary female in the dynasty, and they made great gifts and sacrifices to her. This waqa could be the same one seen today on the esplanade in front of the Qorikancha.

Summary

This chapter featured only three of Qosqos numerous seq'es and many of their waqas but from this you should get a good sense of the intricate roles that seq'es and waqas served in Inka culture. The derivations of names were described, as well as were locations and functions. The seq'e system was central to Inka culture, and to daily life with the ayllus and panaqa members assigned to care for and honor the waqas on the seq'es of their responsibility. Now that seq'es and waqas have been explored, you will be able to gain a much better understanding of the Inka astronomy that is discussed in the next chapter regarding constellations.

9

Inka Constellations

Contents

Types of Inka Constellations	153
Principal Inka Constellations	153
Qollca	154
Urquchillay and Qatachillay	155
Dark Constellations	155
Ch'aska Punchu, Chacra, Orqorara, Chakana, Quntur, Suyuntuy, and Huaman	159
Chuchuqoyllor	160
Mallki	161
Kotu Sankha	161
Puma Yunta	161
Laja Haykuna	161
Summary	161

This chapter explores some of the principal astronomical constellations of the Inkas. Information about these constellations was gathered from Spanish chronicals and frrom the oral traditions of Quechua speakers in the Coordillera de los Andes, primarily at locations such as Qosqo, Apurimaq, Ayacucho, Puno, the Altiplano. Also, the Sacred Valley, Qurawasi, and Paucartambo which are the places where author Milton's grandparents and parents were born and the area where he lives today. Milton learned about many of the constellations through oral traditions that were passed down to him before he ever read about them. The chronicals say where some of the constellations are in the sky. But by growing up here, by knowing the traditions, and by

---

The original version of the chapter has been revised. A correction to this chapter can be found at https://doi.org/10.1007/978-3-031-67580-5_12

© The Author(s), under exclusive license to Springer Nature Switzerland AG 2024, corrected publication 2025
S. Gullberg, M. Rojas Gamarra, *Inca Cosmovision*, Astronomers' Universe, https://doi.org/10.1007/978-3-031-67580-5_9

Fig. 9.1 Machu Piqchu Night Sky. (Watercolor printed with permission. © 2024, Jessica Gullberg. All rights reserved)

understanding and experiencing the worldview of the Inkas, Milton has gathered information otherwise lost due to the Spaniards (Fig. 9.1).

Humans throughout history have wondered what the bright points in the sky meant. The Inkas knew these stars and they tried to order them. Their classifications included not only stellar constellations but also images in the dark regions of the sky. Each cosmological figure had a specific meaning; living beings had representatives in the sky that were responsible for them and their procreation. For example, Yaqana, the llama constellation, had responsibility for all auquenides (llama, vicuña, and alpaca). Inka constellations in the sky include such as Mach'aqway, Yaqana, Laja manta, Collqa, and Chaska punchu. The importance of the constellations in Inka culture are described, including legends related to them.

From the time of Manqo Qapaq to that of Atawallpa, the Inkas felt a great respect for nature and the universe. They thought they were the children of the stars and that is why they put names to individual stars and regions in the night sky. They and their predecessors studied them for many years and were able to make a calendar that was based upon the cycles of Inti the Sun, the stars, and constellations. Their calendar allowed them to determine dates to plant and harvest different products such as corn, potatoes, cassava, oqa, coca, peanuts, sweet potatoes, quiwicha, and quinoa. By observing the heliacal rise of Qollqa, the Inkas were able to predict if the next year would be good for agriculture. They also followed the Sun's position on the horizon to know when to celebrate various ceremonies associated with solstices and equinoxes. Among their constellations were Qollqa, Urquchillay and Catachillay, Chuquichinchay, Choquechinchay, and Chaska Waraqa Mach'aqway, the flame in the eyes of llamaqñawin we know as Alpha and Beta Centauri. Others were Mayu, Yaqana and Yutu, Quntur, Suyuntu and Huaman, Orqorara and Ranch, Foot, Chasqa Punchu, Chaqana, Chaska, Aquayoq and Catuilla, Chuchuqoyllor, Mallki Kotu Sankha, Puma Yoke, Flagstone Hayquna, and Kapuwara Wara. Several of these were depicted on a plate of gold in the Qorikancha (see Fig. 10.4).

Types of Inka Constellations

There are three main types of Inka constellations: (1) constellations of individual stars or clusters, for example the Chuchuqoyllor constellation that is two stars together for Qollqa, (2) constellations that bind different stars such as Ch'aska Waraqa, which is the tail of the Scorpion, and (3) constellations formed by bright and dark areas in the sky, such as Laja Haykuna. There also are mixed constellations between such as Yaqana, the llama, that is composed by a dark area and two bright stars that are called Llamaqñawin, found in the contemporary Southern Cross.

Principal Inka Constellations

In Andean worldview there are philosophies deeply rooted in the way of life and two of them regard the concepts of paqarina and waqa. Paqarina is a place where the Andean people thought their ancestors originated and this generally was located near their homes or within sight from where they lived or walked. A waqa is a place or thing for which they had great respect and held as Apus (gods). These often coincided with the paqarina, but could also be objects, animals, plants, and/or intangible or phenomenological entities, in general anything that caused them admiration, respect, and joy, whether or not logically explainable. Many terrestrial waqas were represented in the sky. Also, animals or plants that were not considered waqas could as well have associated constellations; these constellations were responsible for care of the waqas that they represented. Those of animals and plants were responsible for procreation and for abundant food. Constellations also could be for protection, for example the Mach'aqway constellation was asked to keep the Yaqumama from hurting them when at the Amazon.

Qollca

Qollqa was the most important constellation of the Inkas, and they envisioned it in the open cluster known as the Pleiades. In English Qollqa means barn and is where different dry products were stored such as corn, beans, quinoa, and quiwicha. The qollqa were important because the people knew that if needed, these reserves were available. The constellation that represented these sites was Qollqa and it was closely linked to food and life. It was considered as the Paqarina of all other constellations.

Qollqa was a granary, and at first glance it can look like a whole grain in the sky (see Fig. 9.2), so it was linked to agriculture; this constellation has its heliacal rise near the winter solstice in June. This is the beginning of winter and related freezing and for this reason it was also sometimes called Onqoy, which means disease.

Every morning there was a person in the Qorikancha who was responsible for observing when Qollqa appeared on a hill of Qosqo with an angle of inclination of approximately 20 degrees to the horizontal plane. When it was first seen, the Inkas began the year; this period coincided approximately with the June solstice and a festival called Intirraymi was celebrated. Qollqa was very important and thus was included in the drawing of Pachaquti Salqamaywa (see Fig. 9.3).

Fig. 9.2 Hubble Refines Distance to the Pleiades Star Cluster (Qollqa). (Reprinted from Hubblesite (https://hubblesite.org/contents/media/images/2004/20/1562-Image.html?Topic=104-stars-and-nebulas&keyword=pleiades—Accessed on the 21st of June 2024). © 1996, NASA, ESA and AURA/Caltech. Public domain)

Urquchillay and Qatachillay

These constellations were located in the present-day constellation of Lyra, and within it are three important stars: Vega, Altair, and Deneb. Urquchillay represented lambs and therefore was responsible for their care and their procreation. It is thought that Urquchillay was Vega and Qatachillay was Deneb, and these also appear in the drawing of Salqamaywa in Fig. 9.3. In Qosqo and Qurawasi these were called crossover stars because they represented stones used to cross a river.

Dark Constellations

The river in the sky, Ch'askamayu, is the Milky Way. The Inkas saw dark constellations of great cosmological significance in parts of the Milky Way where areas of interstellar gas and dusk block the light of the stars behind them.

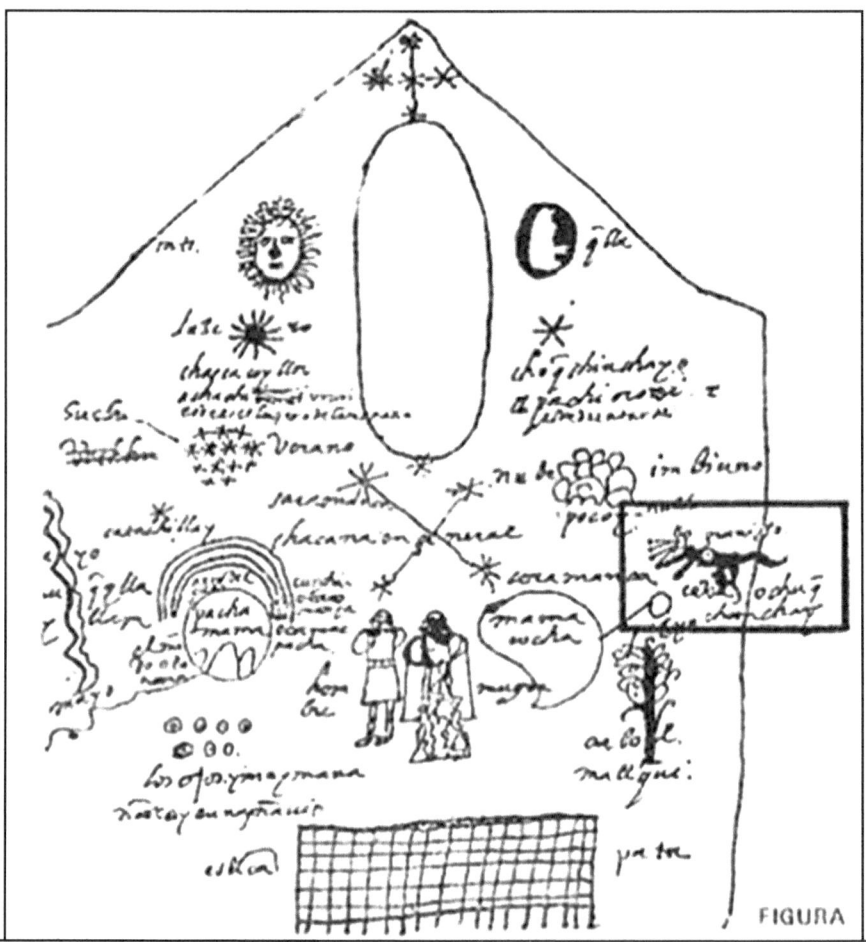

Fig. 9.3 The Qollqa constellation is on the lower left. (Reprinted with permission from (Magli, 2004). © 2005, Birkhaeuser Verlag AG. All rights reserved)

Following are descriptions of the seven Inka dark constellations, from right to left, and they were observed during author Steven's field research at Llaqtapata (Gullberg, 2020; Gullberg et al., 2020).

1. *Mach'aqway*

This constellation represented snakes in general and especially in Antisuyu since Yaqumama, or Anaconda, live there. The Antisuyu is the region of jungle where the Amazon begins. The runas asked this constellation to keep them from being bitten by snakes when they were in the jungle. This also was the representative of the terrestrial Mach'aqway that was the Apu of Ukhupacha, the Apu of wisdom and of knowledge.

Fig. 9.4 Inka dark constellations of the milky way—the procession from right to left: (1) Mach'aqway the serpent, (2) Hanp'atu the toad, (3) Yuthu the tinamou, (4) Yaqana the mother llama, (5) Uñallamacha the baby llama, (6) Atoq the fox, and (7) Michij runa the shepherd; as a point of reference (3) Yutu is also the Coalsack. (Watercolor inspired by Gary Urton and Miguel Araoz Cartagena and printed with permission. © 2024, Jessica Gullberg. All rights reserved)

Mach'aqway, the serpent, leads this dark celestial procession as the constellations gradually proceed right to left across the night sky. Mach'aqway is viewed at the beginning of the rainy season. The serpent's dark figure is long like a snake and travels head before the tail (Urton, 1981).

2. *Hanp'atu*

Hanp'atu, the toad, follows closely behind Mach'aqway. Toads were thought of as bad omens because they were created by the devil. Hanp'atu is a much smaller dark section of the Milky Way to the left of the snake (Urton, 1981).

3. *Yuthu*

Tinamous are birds indigenous to the Andes and are of very ancient lineage. Yuthu, the tinamou, follows Hanp'atu in the Milky Way and likewise is much smaller than Mach'aqway. This Yuthu is adjacent to the constellation of the Southern Cross, is at zenith on the December solstice, and at nadir on the June solstice. Yuthu is known as the *Coalsack* in Western astronomy (Urton, 1981).

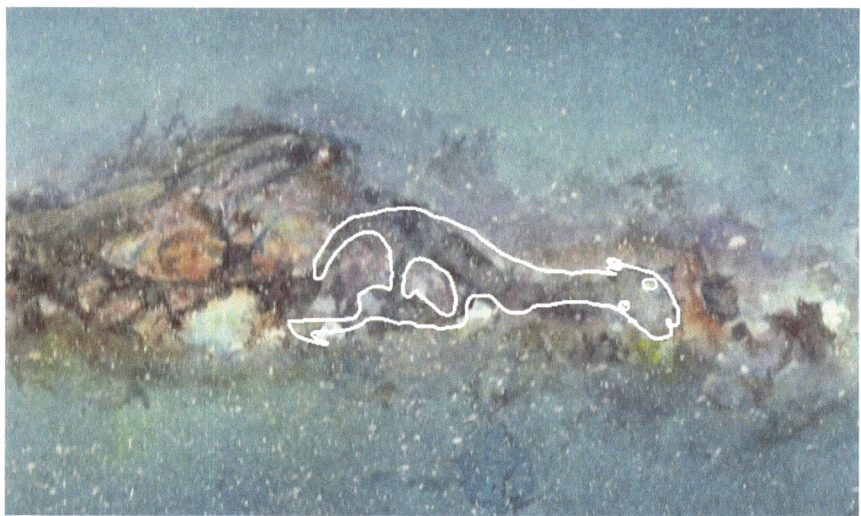

Fig. 9.5 Yaqana and Uñallamacha. Detail taken from Fig. 9.4. Watercolor inspired by Gary Urton and Miguel Araoz Cartagena and printed with permission. © 2024, Jessica Gullberg. All rights reserved

4. **Yaqana**

According to oral traditions in the Qosqo area, Yaqana's eyes (llamaqñawin) were Alpha and Beta Centauri, and Yaqana (a llama) with a long neck is swimming in a river of stars. It is easy to look at the sky and see the constellation, and also see the cria (baby llama) that breastfeeding (see Fig. 9.5).

Llamas figure prominently in many aspects of Inka culture and this celestial figure was thought to animate llamas on the Earth (Salomon & Urioste 1991). Yaqana is a constellation much larger than Hanp'atu or Yuthu and it dominates the Inka dark constellation section of the Milky Way. Yaqana is situated between Centaurus and Scorpio. The prominent stars α and β Centauri serve as the llama's eyes and as such are known as *Llamacñawin*, the "eyes of the llama" (Urton, 1981).

5. **Uñallamacha**

Below Yaqana is a smaller dark constellation called Uñallamacha that is said to be a cria, a baby llama, suckling its mother (Urton, 1981; Fig. 9.5).

6. **Atoq**

Following Yaqana and Uñallamacha is the somewhat smaller constellation of Atoq, the fox. Atoq lies on the ecliptic between the constellations of Scorpio and

Sagittarius and the Sun enters it during the December solstice. Urton (1981) relates that the Milky Way and Atoq catch up and rise with the summer solstice Sun in the southeast during the same period of time that terrestrial baby foxes typically are born. "… the sun rises into [Atoq] … from about December 15 to December 23" (Urton, 1981, p. 70).

7. **Michij runa**

While Urton describes a second Yutu after Atoq, a painting by Miguel Araoz Cartagena of the dark constellations that is displayed within the Qorikancha instead includes a shepherd, Michij runa, herding the pantheon across the sky.

Ch'aska Punchu, Chacra, Orqorara, Chakana, Quntur, Suyuntuy, and Huaman

Stars in the constellation Orion represent the Three Marias—Mintaka, Alnilamy, and Alnitak in the belt, and as well Betelgeuse, Bellatrix, Rigel, and Saiph with their own interpretations and mythologies. For example, these all form the constellation of *Ch'aska Punchu*, which in English means Poncho of Stars. The poncho is an article of clothing that is widely used in the Andes due to cold weather at altitude. It has a rectangular shape with an opening in the middle, similar to the unku of the Inkas, only that the latter was closed by the sides and had an orifice for arms.

Another interpretation for this rectangle was that of a *Chacra* which is a cultivation area. A man had a topo (mole) of land for crops and when married he was given a topo more and another half topo for his wife. When they married a Minka took place in which people gathered to build the couple's new home and also to help prepare their land for planting by using Chaquitaqllas. The Three Marias are also called *Orqorara*, which is translated as three stars all the same. The two stars perpendicular to these three are Betelgeuse and Rigel. The Three Marias are in the drawing of Salqamaywa in Figs. 9.3 and 9.6. The stars of Orion's belt have Quechua names: Mintaka was *Quntur*, Alnilam was *Suyuntuy*, and Alnitak was called *Huaman*.

The three Marias with Betelgeuse and Rigel were also called *Ch'aka Tinkucheq*, which in English means a bridge that links. This is because across the Celestial Equator is a bridge linking the two hemispheres with Betelgeuse to the north and Rigel to the south. The Inkas may have been generally familiar with the Celestial Equator and this is implied by their constructing windows that pointed to the celestial south pole.

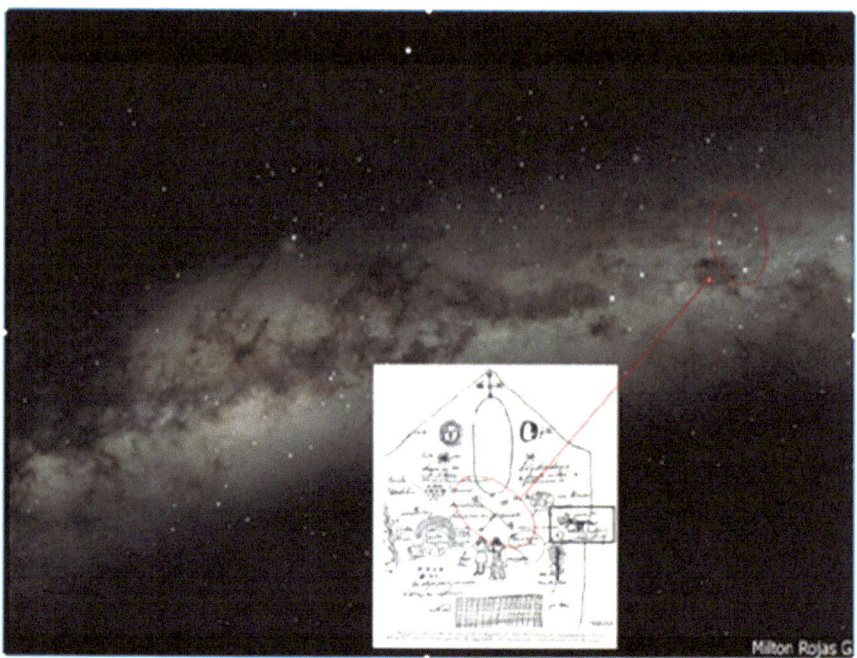

Fig. 9.6 The Ch'akana constellation we know as the Southern Cross. It is included in the Salqamaywa drawing, where can be seen the two stars of the main part of the Southern Cross, one is called Saramanqa (Pot of Maiz) and the other is called Qoqamanqa (Pot of Qoqa); the red arrow points to the Southern Cross. (Modified with permission from from (Magli, 2004). © 2005, Birkhaeuser Verlag AG. All rights reserved)

The constellation called Ch'akana was at the Southern Cross and it also appears on the image of Salkamaywa written as Chacana (see Fig. 9.3). Maiz (corn) and Qoqa (coca) were important in the Inkas' diet and had representative stars in the Southern Cross, one called *Saramanqa* (pot of maiz) and the other *Qoqamanqa* (pot of qoqa).

Chuchuqoyllor

This constellation represented twins. Chuchu also means very dry or hard, arnd these are two stars near to Qollqa that are the feet of Perseus. This constellation was the waqa of people and animals that were born different from normal.

Mallki

This is the constellation of the tree, located in Aries, and can be seen on the horizon. It is also depicted in the Salqamaywa diagram (see Fig. 9.3).

Kotu Sankha

In English, this means coals of fire and is located in the constellation Hyades. Aldebaran is the fire and other nearby stars are the coals.

Puma Yunta

Puma Yunta is a pair of puma friends and is located in Gemini. The eyes of the puma are the stars Castor and Pollux. In Inka mythology the Puma, the Quntur (Condor), and the serpent Mach'aqway were considered sacred. The Quntur was the Apu of Peace and represented the Hananpacha (the space and time of the overworld, the world above), the Puma was the Apu of the Force and represented the Kaypacha (space and time of this world that we perceive with our senses) and the Mach'aqway was the Apu of wisdom and knowledge and represented the Ukhupacha (the space and time of the underworld).

Laja Haykuna

This means entry into the darkness and is located in the constellation of Argo Navis, above of the May Cross. This was like a door to the Chinqana, a tunnel. This was a Paqarina.

Summary

In conclusion, this has been an overview of some of the most important constellations of the Inkas, however there are many more. These constellations are still seen by many rural Andeans but, in part thanks to the Spaniards, they have been long forgotten by others. Information presented here was obtained mainly by interviewing local peoples, many being grandparents living in the provinces of Qosqo and surrounding areas. A great debt is owed to Amauta Emilio Huaman Huillca and his knowledge.

Principles, including Mit'a and Minka, are based on Ayni. Mit'a and Minka are two ways of work organization, and this worldview is also reflected in the constellations. The Inkas believed that the constellations took care of and protected the animals or plants which they represented in accordance with the principle of Ayni. One of the main reasons the Inkas built temples and performed ceremonies was to worship these constellations.

It is important to record and disseminate this knowledge that is being lost, because with it we can preserve the worldview, philosophy, and lifestyle of the Inkas.

10

Inka Astronomy in Daily Life

Contents

Daily Life and Routine	164
Everyday Knowledge	164
Astronomy in the Daily Life of the Inkas	165
Examples of Astronomical Ceremonies and Participation of the People	165
Example of Archaeological Places and Frequency of Visits of the Town	169
Summary	171

Largely from the oral traditions, this chapter explores how astronomy was present in Inka daily life, and not only for the elite but also the non-elite as well. Descriptions of daily life that include astronomy follow, along with descriptions of various aspects of Inka astronomy. Knowledge of the stars and their movements was part of the lives of all Inka people and not just that of the elite. Everyone could see what was in in the sky.

Humans have always been curious to know what was happening in the world in which they lived. Many explanations have been given, initially supernatural, later with mysticism, and eventually those of science. The *Hananpacha*, the name for the sky in the Tawantinsuyu which translates as the universe or world above or space-time above, was a reason for curiosity, reverence, and reciprocity in Inka worldview. It was practical and useful to such an extent that it had to be present in daily life and culture.

There are two ways to understand this worldview—one is to live it and the other is to imagine it. This chapter helps with the latter because is has become common for those who do not live in the countryside to embrace the Western point of view. To understand astronomy in the daily life of the Inkas, it must approached from the "Pacha," space and time.

The original version of the chapter has been revised. A correction to this chapter can be found at https://doi.org/10.1007/978-3-031-67580-5_12

Astronomy was present in everyday life in the Tawantinsuyo. As stated, this knowledge was not elitist and, because of the life principles of the Inkas, it had to be this way.

Word ambiguities can lead to misunderstandings. Scholars as much as possible need to use the words of original languages, because trying to truly define or explain situations, feelings, and ideas with words of a different culture is difficult, and sometimes not possible. Astronomy was present in the daily life of common people in the Llaqta in the time of the Inkas and where possible Quechua words and meanings are used here for greater insight and clarity.

Daily Life and Routine

In everyday life, what matters is time, so what was done was repeated in cycles of more or less exact time. "Everyday" in the time of the Inkas could be measured by observing the constellation Chaska each day in its heliacal rise, and thus this cycle represented 1 day. It could also be part of daily life to perform ceremonies corresponding to the Pachamama before planting and harvesting—here the cycle could be 2 months, 3 months, or more depending on the product, and the ceremonies that were made for the solstices and equinoxes could also be part of daily life. Daily life depended first on cycles in time, and it had to repeat itself in time. Routine also depends on space—it is a spatial cycle. For example, it was routine in the time of the Inkas to go to Saqsaywaman to perform ceremonies for the Apu Inti.

So, a necessary condition for the existence of daily life and routine is that it must be repeated, and the necessary condition for daily life is in time and for the routine it is in space. All that happened, daily or not, routine or not, in the life of the people in the Tawantisuyo was called *allin kausay* (good living).

Everyday Knowledge

Reality is what can be perceived with senses, and everyday knowledge is objective and normative. Scientific knowledge, and especially astronomical knowledge, was part of the community in the time of the Inkas, especially the affirmation of the normative, because there were many norms in the Tawantinsuyo that were the product of observations of the sky. Therefore, everyday knowledge was the sum of knowledge about reality used in everyday life in a most heterogeneous way, and collective scientific knowledge was the sum of that used in everyday life.

Astronomy in the Daily Life of the Inkas

The cosmos is accessible to our senses, and since man becomes aware that he is immersed in the cosmos, he tries to organize his life with respect to what happens in his environment. In the past what happened in the the cosmos was very much in the consciousness of people. This was for various reasons, because there was no electric light, because they lived in rural environments, and because they were directly dependant upon the environment, especially for food.

When cultures became aware of the cosmos they tried to interpret it with myths, and through this the people learned astronomy. In the Tawantinsuyo and throughout the Andean world pre-Inka peoples tried to harmonize daily life with what was happening in the cosmos. Therefore they learned the sky and this became part of their cultures. In the case of the Inkas, there exist many examples where these orientations can be seen. There are many extant calendars, and even the designs of Llaqta (towns) could have been oriented somewhat to assist calendars through their seq'es (Zuidema, 1964, 1977).

Examples of Astronomical Ceremonies and Participation of the People

The Inkas, as in many cultures, tried to interpret the sky and its different astronomical and meteorological phenomena. They realized that apparent movements of the Sun on the horizon could help predict the type of weather they would experience in different seasons. This became very important for them, because by predicting the weather they were able to have the best harvests (see Fig. 10.1).

For interpreting apparent movements of the Sun, especially on the horizon, it was realized that solar horizon positions repeated in an annual cycle, which they called *Intipwatan* (year of the Sun). It also was known that there were 2 days each year in which a vertical pole cast no shadow at local noon. When the Sun's horizon travel reached its northern extreme in June, the June solstice, the Inkas celebrated with a ceremony they called Intirraimy, the Festival of the Sun.

In the days approaching the solstice another astronomical phenomenon occurred, the heliacal rise of the constellation Qollqa (see Fig. 9.2). Qollqa was the most important constellation of the Inkas, and it is also known as the Pleiades, or Las Pléyades, which in Spanish means granary. Qollqa were places

Fig. 10.1 Sunrise positions on the Solstices and Equinoxes as seen from the Qolqampata, an old Inka neighborhood

where different dry products such as corn or sara, broad beans, quinoa, and quiwicha were stored.

Qollqa were important because they held reserves for times of drought. Qollqa was the constellation that represented these places, and it became intimately linked to food and life and was considered the Paqarina of the other constellations.

There was a person in charge of observing the horizon each morning until Qollqa first appeared after its journey behind the Sun. At the moment of first appearance on the horizon at dawn, the heliacal rise, the person responsible was Kawsaypacha Qamayoq, the Inka Astronomer, and he was aided by his assistants. In a drawing made by the chronicler Guaman Poma (Fig. 10.2), this astronomer is called in old Spanish, Astrologo Pveta (Astrologer Poet that Saves). Figure 10.3 shows author Milton as Kawsaypacha Qamayoq at the time of Intirrami.

The Kawsaypacha Qamayoq had to observe if Qollqa looked dim or bright upon its heliacal rise and he used what he saw to predict if the year would be a good one or a bad one for crops. Essentially he was identifying the effects of an El Niño year and related drought.

The constellations of Qollqa, Chaskaqoyllor (Venus), and the Qoyllorquna (stars) have a special room dedicated to them in the Qorikancha. There is a

10 Inka Astronomy in Daily Life 167

Fig. 10.2 Drawing by Guaman Poma de Ayala, for, Kawsaypacha Qamayoq, the Inka Astronomer whom he called 'Astrologer Poet that Saves'. (Reprinted with permission from (Guaman Poma de Ayala, 2009). © 2010, University Texas Press. All rights reserved)

Fig. 10.3 North wall of the enclosure dedicated to the Qollqa in the Qorikancha. This wall and the alley are aligned to the first heliacal rise of Qollqa and to the June solstice sunrise The photo shows author Milton as Kawsaypacha Qamayoq

window there that directly overlooks the June solstice and it is thought that the heliacal rise of Qollqa was observed there too (see Fig. 10.5).

The Intirraymi ceremony began in the Qorikancha and then moved to the main square. From there it went to Saqsaywaman where a ceremony offered the sacrifice of a llama.

In the square while waiting for sunrise, the attendees together with the Orejones sang songs, moving one foot to the beat, singing and sending ceremonial blown kisses to the Sun.

Because of the importance of this ceremony, it was celebrated throughout the Tawantinsuyu. As such, the exact date had to be known everywhere. Most places had an Intiwatana to know exactly when this phenomenon would occur so they could perform the ceremonies. There also were other ways to communicate this event, such as with two adobe or stone parallel walls that were close, one with long windows and closed on the sides. Inside a fire was set that could be seen through the windows from hill to hill and a message could be sent by covering parts of the windows in a manner that used pre-established codes.

This ceremony was practiced with great awareness and devotion by the Inkas, but was diminished after the arrival of the Spanish. Fortunately, today these ceremonies are still celebrated by the original peoples in the different countries in which the Tawantinsuyu existed.

The Intirraymi ceremony was and still is held on June 24th. After the Spanish arrived this festival was replaced by the festival of Corpus Christi

Fig. 10.4 The replica that is in the Qorikancha of the depiction that Pachakuteq Salqamaywa made with the the figures as told to him and as he interpreted them in approximation. The constellation of Qollqa is on the left side and including it was logical due to its importance

because its celebration fell around the same dates. In the Catholic ceremony, the different saints of the parishes of Qosqo were carried on a litter with a band of musicians and with dancers following behind. In the time of the Inkas the different mummies were carried on litters past those at the Qorikancha (see Figs. 10.6, and 10.7).

Example of Archaeological Places and Frequency of Visits of the Town

Inka communities, like Qosqo, were surrounded by sacred objects and ceremonial places called waqas. These places had a transcendental role in the life of the Indigenous peoples who occupied them: "The Andean landscape is imbued

Fig. 10.5 Window in the Qorikancha facing the sunrise on the June Solstice that also looks at the heliacal rise of Qollqa; this was a very important window and thus was lined with gold sheets, within them there were incrustations of precious stones, that upon the arrival of the Spaniards were stolen and turned into gold bars

Fig. 10.6 Saints were brought from the different temples of Qosqo to the cathedral and then passed in procession through the Plaza de Armas in Corpus Christi. The mummies were removed by the extirpators of idolatry

Fig. 10.7 Dances of Paucartambino—today there are many dances that are taken from Paucartambino folklore to accompany the different Saints in the procession; there are more than 30 varieties of dances

with sacredness. Human destinies are determined in part by the powers of the spirits of the mountains, rocks, springs, rivers, and other topographical features, and generalized in the Pachamama, the earthly matrix" (Bauer & Stanish, 2001). Thus Saqsaywaman, Pisaq, Machu Piqchu, Ollantaytambo, Piquillaqta, Tambomachay, and natural places like Huamanqocha, Ankasqocha, the apu Salqantay, and Waqa 44 were revered, some because they were considered as places of origin (paqarina), others because they were Apu, and more because they were beautiful. They were chosen as paqarina or waqa and it must be remembered that the pachamama, the waqas, and the apus all lived for the people of the Coordillera de los Andes. This is why those in the Tawantinsuyo had to constantly communicate with these waqas. And this is why Amauta Emilio Huaman Huillca goes to Waqa 44 many times a year, because it was there that he discovered alignments for the solstices and equinoxes.

Summary

In conclusion, all the people in the time of the Inkas had knowledge of astronomy and not just the elite for the following reasons:

1. They had more contact with nature.
2. At night it is more evident that we are immersed in the cosmos and in the time of the Inkas there was no light pollution, therefore phenomena such as lunar eclipses, meteorites, and comets were quite visible.
3. Many of the activities that were carried out were directly related to astronomical phenomena, the main activity being the festival of Intirraymi that was held at the time of the June solstice. These activities were not only carried out in Qosqo but throughout the empire, for which many solar observatories existed.

4. The first heliacal rise of the Qollqa constellation that marked the beginning of the new Inka year occurred just prior to the time of the June solstice; the two joint phenomena were present for the beginning of the allin watan (good year). All the ayllus knew this phenomenon.
5. The constellations were directly related to their corresponding terrestrial animals and plants, each animal and plant had a counterapart in the sky that was in charge of protecting them and their reproduction, even the Urubamba river itself had a counterpart which was the Milky Way, and this was called *Ch'askamayu* (river of stars).
6. They had to know the seasons, because most were farmers, but also for hunting such as tarucas, ducks, and rabbits.
7. Ceremonies were carried out simultaneously throughout the Tawantinsuyo, which is why sundials have been found in many archaeological complexes.
8. The Apus, Waqas and Paqarinas were reason for routine, the people constantly traveled to these places that they venerated because they were so highly respected.
9. Mystical ceremonies, agriculture, and education were the reason for the existence of daily life.
10. In order for crop sowing to be at precise times, it was necessary to know the movements of the Sun.
11. Astronomically oriented constructions were public places where ceremonies were held in which the whole town participated.
12. In order to carry out ceremonies on precise days throughout the Tawantinsuyo, the town and especially the Quraqa in charge had to be aware of what was happening in the cosmos, for example the heliacal rise of a star or the position of the shadow in an Intiwatana (sundial).
13. The existence of Intiwatanas in many places demonstrated the need to measure time and to know the cycles that the Pachamama had in its astronomical movement for the aforementioned purposes. What stands out is that these clocks and even calendars were not only in Qosqo but in all the Tawantinsuyo.

The science of astronomy was present in the daily lives of all Inkas. Not only astronomy, but all science and all knowledge was not elitist, it was a part of daily knowledge for everyone.

11

Conclusion

Contents
Final Thoughts 174

The Inkas in the Tawantinsuyo created a great culture with beliefs, customs, habits, and ways of being. They developed a worldview that included fundamental principles of life such as Ayni (reciprocity) and Kawsaypacha (everything lives). Both, for the Inkas, came as natural laws.

The ancient Inkas practiced horizon astronomy in the Coordillera de los Andes through positional observations of the rising and setting Sun on days of ceremonial and agricultural significance. They also observed rises for constellations such as Qollqa. A majority of extant sites with astronomical orientations display attention to the solstices.

The pillars above Q'espiwanka lie on a sightline to the June solstice sunrise when viewed from a large boulder at the center of the coutyard of the palace below. The Sacred Plaza of Machu Piqchu, the River Intihuatana, and the Llaqtapata Sun Temple lie approximately along an axis established by the June solstice sunrise and December solstice sunset. The carved rock waqa at Q'enqo Grande is rich with astronomical orientations and its neighbor at Lacco features even more. Waqa 44, with cylinders designed to indicate orientations for cardinal solar horizon events, would have been used to determine where to view solsticial and equinoctial sunrises and sunsets.

The Inkas organized their culture with seq'es and waqas and these fit into their cosmology and astronomy. 41 seq'es surrounded Qosqo, each with numerous waqas. Waqas were sacred entities for the Inkas, and today they are somewhat compared with shrines. They were deified elements and were worshipped. As with Kawsaypacha, they all were alive and thus required communication from and care by the Inkas. Waqas could be such as springs, rocks,

Fig. 11.1 Author Milton and his research assistant sons in the field. (Published with permission by Milton Rojas Gamarra)

trees, caves, t'oqos, palaces, Apus, rivers, and anything though to be sentient and powerful.

Seq'es were the Inka organizational systems surrounding settlements. The one we know most about is the one around the Inkas' primary Llaqta of Qosqo where it is said that 41 seq'es existed and with them were more than 350 waqas, the majority destroyed during the Spanish extirpation of idolatries. The waqas on the seq'es were cared for by the panaqas and ayllus. Certain seq'es and waqas are thought to have had astronomical significance. The first three seq'es of Chinchaysuyo and many of their waqas were used as examples of the complexity and significance of this part of Inka culture.

Many fascinating cosmological Inka contellations were featured—those that were stellar, dark, and hybrid. And then was shared how astronomy fit into the daily lives of all Inkas, regardless of social status (Fig. 11.1).

Final Thoughts

Aspects of the Quechua language have been explored in depth and used in an effort to immerse you in a better understanding of Inka culture—what the Inkas believed and why their astronomy and cosmovision developed as it did. Even more importantly, though, and the central premise of this book, is what can be learned from Inka oral traditions. The Inkas did not have a written language and passed on remembrance orally from generation to generation. Even though the empire has long since gone, transmission of these traditions orally never stopped. Author Milton, who is Quechua, has been a beneficiary of these recollections and we have used them here to describe Inka astronomy and cosmovision in a way that has not been published before for the world to see. And in this fashion, we are preserving these traditions before they are lost.

A picture emerges of a culture interwoven with cosmology and astronomy. The Inkas possessed sophisticated astral knowledge and as solar worshippers they chose to incorporate and display alignments and features of the Sun, their god, in their many temples and waqas throughout the empire. These waqas point to a culture that was both interwoven with the Sun and that possessed the technical ability to accurately encode their celestial knowledge into any structure or carving they so desired.

The intent of this book is to provide a greater insight into Inka Cosmovision and Astronomy. This we hope to have accomplished though a heightened sense of Inka culture by describing it with Quechua words and also by the incomparable insight we have been able to use from the Inka oral traditions passed through the generations and ultimately to author Milton. We hope you have enjoyed what we have been able to share and that it will be valuable to you in the future.

Correction to: Inca Cosmovision: The Astronomical Legacy of an Andean Empire

Correction to:
S. Gullberg, M. Rojas Gamarra, *Inca Cosmovision*, Astronomers' Universe,
https://doi.org/10.1007/978-3-031-67580-5

The original version of this book was inadvertently published with few errors in the following Chapters 4, 6, and 8–10. The correction chapters and the book have been updated with these changes.

The updated version of these chapters can be found at
https://doi.org/10.1007/978-3-031-67580-5_4
https://doi.org/10.1007/978-3-031-67580-5_6
https://doi.org/10.1007/978-3-031-67580-5_8
https://doi.org/10.1007/978-3-031-67580-5_9
https://doi.org/10.1007/978-3-031-67580-5_10

Bibliography

Aranibar, C. (Ed.). (2017 [1613]). *Nueva cronica buen gobierno. Ministerio de Relaciones Exteriores: Biblioteca Nacional del Peru*. The Royal Danish Library.

Arriaga, P. J. (1968 [1621]). *The extirpation of idolatry in Peru*. In L. C. Keating (Ed. & Trans.). University of Kentucky Press.

Aveni, A. (1981a). Horizon astronomy in Incaic Cusco. In R. Williamson (Ed.), *Archaeoastronomy in the Americas* (pp. 305–318). Ballena Press.

Aveni, A. (1981b). Tropical Archaeoastronomy. *Science, 213*(4504), 161–171.

Bauer, B. (1998). *The sacred landscape of the Inca: The Cusco Ceque system*. University of Texas Press.

Bauer, B., & Dearborn, D. (1995). *Astronomy and empire in the ancient Andes: The cultural origins of Inca sky watching*. University of Texas Press.

Bauer, B., & Stanish, C. (2001). *Ritual and pilgrimage in the ancient Andes*. University of Texas Press.

Benson, E. P., & Cook, A. G. (2001). *Ritual sacrifice in ancient Peru*. University of Texas Press.

Betanzos, J., (1996 [1576]). *Narrative of the Incas*. Buchanan, D. (Ed.) and (R. Hamilton, Trans.). University of Texas Press.

Burger, R. L. (1992). *Chavin and the origins of Andean civilization*. Thames and Hudson.

Christie, J. J. (2007). Did the Inka copy Cusco? An answer derived from an architectural-sculptural model. *Journal of Latin American and Caribbean Anthropology, 12*(1), 164–199.

Cieza de León, P., (1998 [1555]). *The discovery and conquest of Peru*. In A. Parma-Cook, & N. Cook (Eds. & Trans.). Duke University Press

Cobo, B. (1983 [1653]). *History of the Inca empire: An account of the Indians' customs and their origin, together with a treatise on Inca legends, history, and social institutions.* In R. Hamilton (Ed. & Trans.). University of Texas Press.

Cobo, B. (1990 [1653]). *Inca religion and customs.* In R. Hamilton (Ed. & Trans.). University of Texas Press.

D'Altroy, T. (2002). *The Incas.* Wiley-Blackwell.

de la Vega, G. (1961 [1609]). *The Royal Commentaries of the Inca.* Avon.

Dearborn, D., & Schreiber, K. (1986). *Here comes the sun: The Cusco-Machu Picchu connection* (pp. 15–36). Archaeoastronomy.

Dearborn, D., & White, R. (1989). Inca observatories: Their relation to the calendar and ritual. In A. Aveni (Ed.), *World Archaeoastronomy* (pp. 462–469). Cambridge University Press.

Dearborn, D., Schreiber, K., & White, R. (1987). Intimachay, a December solstice observatory. *American Antiquity, 52*, 346–352.

DeLeonardis, L., & Lau, G. (2004). Life, death, and ancestors. In H. Silverman (Ed.), *Andean archaeology* (pp. 77–115).

Eliade, M. (1972). *Shamanism: Archaic techniques of ecstasy.* Princeton.

Gamarra, M. R. (2017). *Matemática Inka.* Cusco.

Gamarra, M. R., & Zen Vasconcellos, C. A. (2019). The constellations and space-time concept according to the Inkas. *Astronomische Nachrichten, 340*(1–3), 18–22.

Gamarra, M. R., Estrazulas, M., Gullberg, S. R., & Zen Vasconcellos, C. A. (2024). The Ushnus in the astronomy of the Inca culture. *Astronomiche Nachrichten, 34*(2–3).

Gasparini, G., & Margolies, L. (1980). *Inca architecture.* Indiana University Press.

Guaman Poma de Ayala, F. (2009 [1613]). *The first new Chronicle and good government: On the history of the world and the Incas up to 1615.* In R. Hamilton (Ed. & Trans.). University of Texas Press.

Gullberg, S. (2010). Inca solar orientations in southeastern Peru. *Journal of Cosmology, 9*, 2078–2091.

Gullberg, S. (2015). Marking time in the Inca empire. *Journal of Skyscape Archaeology, 1*(2), 217–241.

Gullberg, S. (2016). Astronomy and the Ceques of Cusco. In F. Silva, J. M. Malville, T. Lomsdalen, & F. Ventura (Eds.), *The materiality of the sky* (pp. 255–266). Sophia Center Press.

Gullberg, S. (2018). Inca astronomy: Horizon, light, and shadow. *Astronomische Nachrichten., 340*(1–3), 23–29.

Gullberg, S. R. (2020). *Astronomy of the Inca empire: Use and significance of the sun and the night sky.* Springer Nature.

Gullberg, S., & Malville, J. M. (2011). The astronomy of Peruvian Huacas. In W. Orchiston, T. Nakamura, & R. Strom (Eds.), *Highlighting the history of astronomy in the Asia-Pacific region.* Springer.

Gullberg, S., & Malville, J. M. (2014). The river Intihuatana: Huaca sanctuary on the Urubamba. *Mediterranean Archaeology & Archaeometry, 14*(3), 179–187.

Gullberg, S., & Malville, J. M. (2017). Caves, liminality, and the sun in the Inca World. *Culture and Cosmos, 21*(1 & 2), 193–214.

Gullberg, S. R., Hamacher, D., Martin Lopez, A., Mejuto, J., Munro, A., & Orchiston, W. (2020). A cultural comparison of dark 'constellations' of the milky way. *Journal of Astronomical History and Heritage, 23*(2), 390–404.

Hemming, J. (1970). *The conquest of the Incas*. Harcourt.

Hemming, J., & Ranney, E. (1982). *Monuments of the Incas*. University of New Mexico Press.

Krupp, E. C. (1983). *Echoes of the ancient skies: The astronomy of lost civilizations*. Harper & Row.

Lee, V. (2000). *Forgotten Vilcabamba*. Sixpac Manco.

Magli, G. (2004). *On the astronomical content of the sacred landscape of Cusco in Inka times*. Dipartimento di Matematica del Politecnico di Milano. https://www.researchgate.net/publication/2171124_On_the_astronomical_content_of_the_sacred_landscape_of_Cusco_in_Inka_times/figures?lo=1

Malville, J. M. (2009). Animating the inanimate: Camay and astronomical Huacas of Peru. In J. A. Rubiño-Martín, J. A. Belmonte, F. Prada, & A. Alberdi (Eds.), *Cosmology across cultures* (ASP conference series, 409) (pp. 261–266). Astronomical Society of the Pacific.

Malville, J. M., Thomson, H., & Ziegler, G. (2006). The sun Temple of Llactapata and the ceremonial neighborhood of Machu Picchu. In T. Bostwick & B. Bates (Eds.), *Viewing the sky through past and present cultures* (pp. 327–339). City of Phoenix Parks, Recreation and Library.

Malville, J. M., Zawaski, M., & Gullberg, S. (2008). Cosmological motifs of Peruvian Huacas. In J. Vaiškūnas (Ed.), *Astronomy and cosmology in folk traditions and cultural heritage* (*Archaeologia Baltica*, 10) (pp. 175–182). Vilnius.

McPhee, R. (2022, June 25). *The Ancient Pathway Across South America*. Explorersweb. https://explorersweb.com/the-ancient-pathway-across-south-america/

Niles, S. (1987). *Callachaca: Style and status in an Inca community*. University of Iowa Press.

Niles, S. (1999). *The shape of Inca history: Narrative and architecture in an Andean empire*. University of Iowa Press.

Orlove, B. S., Chiang, J. C., & Cane, M. A. (2000). Forecasting Andean rainfall and crop yield from the influence of El Niño on Pleiades visibility. *Nature, 403*, 68–71.

Paternosto, C. (1989). *The stone and the thread: Andean roots of abstract art*. University of Texas Press.

Pizzaro, P. (1921 [1571]). *Relation of the discovery and conquest of the kingdoms of Peru*. BiblioLife.

Polo de Ondegardo, J. (1965 [1571]). *A report on the basic principles explaining the serious harm which follows when the traditional rights of the Indians are not respected*. (A. Brunel, J. Murra, & S. Muirden, Trans.). Human Relations Area Files.

Reinhard, J. (1985). Sacred Mountains: An ethno-archaeological study of high Andean ruins. *Mountain Research and Development, 5*(4), 299–317.

Reinhard, J. (2002). *Machu Picchu, the sacred center*. Cusco.
Reinhard, J., & Ceruti, C. (2005). Sacred Mountains, ceremonial sites, and human sacrifice among the Incas. *Archaeoastronomy, XIX*, 1–43.
Rowe, J. (1946). *Inca culture at the time of the Spanish conquest*. USGPO.
Rowe, J. (1990). Machu Picchu a la luz de documentos de siglo XVI. *Histórica, 14*(I), 139–154.
Salazar, L. (2004). Machu Picchu: Mysterious Royal Estate in the cloud forest. In L. Burger & L. Salazar (Eds.), *Machu Picchu, unveiling the mystery of the Incas* (pp. 21–48). Yale University Press.
Salazar, F. E. E., & Salazar, E. E. (2014). *Cusco and the Sacred Valley of the Incas* (2nd ed.). Tankar E.I.R.L.
Salomon, F., & Urioste, G. (1991). *Introductory essay in the Huarochiri manuscript: A testament of ancient and colonial Andean religion*. University of Texas Press.
Sarmiento de Gamboa, P. (2009 [1572]). *History of the Inkas*. In B. Bauer, & V. Smith (Trans.). BiblioBazaar.
Sherbondy, J. (1992). Water ideology in Inca Ethnogenesis. In R. V. H. Dover, K. E. Seibold, & J. H. McDowell (Eds.), *Andean cosmologies through time: Persistence and emergence* (pp. 46–66). Indiana University Press.
Squire, E. G. (1878). *Peru: Incidents of travel and exploration in the land of the Incas*. Macmillan and Co.
Staller, J. E. (2008). Dimensions of place: The significance of centers to the development of Andean civilization: An exploration of the *Ushnu* concept. In J. E. Staller (Ed.), *Pre-Columbian landscapes of creation and origin* (pp. 269–314). Springer.
Stastny, F. (1989). El Arte de la Nobleza Inca y La Identidad Andina. In I. Lavin (Ed.), *WORLD ART: Themes of Unity in diversity. Acts of the XXVIth international congress of the history of art* (pp. 731–738). Penn State Press.
Taylor, G. (1974). Camay, camac, et camasca dans le manuscript quechua de Huarochiri. *Journal de la Societe des americanistes, 63*, 231–243.
Tello, J. C. (1943). Discovery of the Chavin culture in Peru. *American Antiquity, 9*(1), 135–160.
Thomson, H. (2001). *The White rock: An exploration of the Inca heartland*. Weidenfeld and Nicolson.
UNESCO World Heritage Centre. (2021). *Chankillo Archaeoastronomical Complex*. https://whc.unesco.org/en/list/1624/
Urton, G. (1981). *At the crossroads of earth and sky: An Andean cosmology*. University of Texas Press.
Urton, G. (2003). *Signs of the Inka quipu: Binary coding in the Andean knotted-string records*. University of Texas Press.
Urton, G. (2011). Tying the archive in knots, or: Dying to get into the archive in ancient Peru. *Journal of the Society of Archivists, 32*(1), 5–20.
Van de Guchte, M. J. D. (1990). *Carving the World: Inca monumental sculpture and landscape*. Doctoral dissertation. University of Illinois.

Wright, K., & Valencia, A. (2000). *Machu Picchu: A civil engineering marvel*. ASCE Press.

Yupanqui, T. C., (2005 [1570]). *An Inca account of the conquest of Peru*. In (R. Bauer Trans.). University Press of Colorado.

Yupanqui, T. C., (2006 [1570]). *History of How the Spaniards Arrived in Peru*. In (C. Julien Trans.). Hackett.

Ziolkowski, M., & Kosciuk, J. (2018). Astronomical observations in the Inca Temple of Coricancha (Cusco)? A critical review of the hypothesis. *TEKA Komisji Architektury, Urbanistyki I Studiów Krajobrazowych, XIV*(1), 7–33.

Zuidema, R. T. (1964). *The Ceque system of Cusco: The social organization of the Capital of the Inca*. E.J. Brill.

Zuidema, R. T. (1977). The Inca calendar. In A. Aveni (Ed.), *Native American astronomy* (pp. 219–259). University of Texas Press.

Zuidema, R. T. (1981). Inca observations of the solar and lunar passages through zenith and anti-zenith at Cuzco. In R. Williamson (Ed.), *Archaeoastronomy in the Americas* (pp. 319–342). Ballena Press.

Zuidema, R. T. (1982). Catachillay: The role of the Pleiades and of the southern cross and α and β Centauri in the calendar of the Incas. In A. Aveni & G. Urton (Eds.), *Ethnoastronomy and Archaeoastronomy in the American tropics* (pp. 203–229). New York Academy of Sciences.

Zuidema, R. T. (1983). Hierarchy and space in Incaic social organization. *Ethnohistory, 30*(2), 49–75.

Zuidema, R. T. (1986). The place of the *Chamay Wariqsa* in the rituals of Cuzco. *Amérindia, 11*.

Zuidema, R. T. (1989). El Ushnu. *Reyes y Guerreros. Ensayos de Cultura Andina*. Grandes Estudios Andinos. Fomciencias, 402–454.

Zuidema, R. T. (1990). *Inca civilization in Cuzco*. University of Texas Press.

Zuidema, R. T. (2005). The astronomical significance of a procession, a pilgrimage and a race in the calendar of Cusco. In J. W. Fountain & R. M. Sinclair (Eds.), *Current studies in Archaeoastronomy: Conversations across time and space* (pp. 353–367). Carolina Academic Press.

Zuidema, R. T. (2007). Solar and lunar observations in the Inca calendar. In C. Ruggles & G. Urton (Eds.), *Skywatching in the ancient World: New perspectives in cultural astronomy* (pp. 269–285). University Press of Colorado.

Zuidema, R. T. (2008a). Pilgrimage and ritual movements in Cuzco and the Inca empire. In J. M. Malville & B. Saraswati (Eds.), *Pilgrimage: Sacred landscapes and self-organized complexity* (pp. 269–288). Indira Ghandi National Centre for the Arts.

Zuidema, R. T. (2008b). The astronomical significance of ritual movements in the calendar of Cuzco. In J. E. Staller (Ed.), *Pre-Columbian landscapes of creation and origin* (pp. 249–268). Springer.

Zuidema, R. T., & Quispe, M. U. (1973). A visit to god—The account and interpretation of a religious experience in the Peruvian community of Choque-Huarcaya. *Bijdragen tot de Taal-, Land- en Volkenkunde, 124*(1), 22–39.

Index

Agricultural terrace, 26
Agriculture, 26
Andagoya, 13
Antisuyu, 22
Archaeoastronomical implements, 63–64
Archaeoastronomy, 59–68
Astronomy, 13, 14, 37, 38, 59, 67, 69, 78, 79, 95, 157, 175
Atahualpa, 12, 13
Ausengate, 19
Aveni, A., 70
Ayllus, 22, 79

Bauer, B., 9, 11, 14, 17, 18, 22, 26, 28, 61, 63, 70, 77, 78
Benson, E.P., 20

Calendar, 23, 67
Camay, 19, 20, 27, 37, 38
Canals, 27
Cañari, 8, 11
Cane, 78
Capac Raymi, 26, 72
Carved rocks, 36
Caves, 16, 18, 19, 79, 82, 84
Cayao, 22
Celestial myths, 5
Celestial river, 76
Celestial sphere, 59–60
Cerro Sayhua, 65, 85
Chachapoya, 8
Chancas, 8, 9, 12, 70
Channels, 27, 92, 94
Chavín, 7
Chiang, J.C., 78
Chimú, 7, 9
Chinchaysuyu, 22
Chinchero, 11
Chullpas, 32
Cibichimpo Rontocay, 12
Colla, 8
Collana, 22
Collasuyu, 22
Conchuco, 8
Condors, 20, 23, 36, 38
Conquistadors, 5

Constellations, 77
Cook, A.G., 20
Coricancha, 70
Cortés, 12
Cosmology, 8, 14, 18, 19, 36, 69–95, 175
Coursed, 31
Coya Cusirimay, 12
Crop management, 61, 67, 70
Cuntisuyu, 22
Cusco, 11, 21, 60, 61, 66, 70, 85, 90

D'Altroy, T., 8, 9
Dark constellations, 77, 95, 156
Dearborn, D., 8, 11, 17, 26, 61, 63, 72, 77, 78
December solstice, 76, 86, 173
de la Vega, G., 24, 49, 54, 57, 62, 79, 97, 98, 135, 136, 148, 155
DeLeonardis, L., 20

Ecliptic, 61, 62, 93, 158
Eliade, M., 21
El Niño, 26, 78
Equinoxes, 61–63, 67, 77, 173

Fountains, 27
Francisco Pizarro, 9

Gasparini, G., 21, 29, 31, 32, 36
Gnomons, 11, 62, 67, 79

Hanan, 21
Hanan Pacha, 18
Hemming, J., 8–14, 16–18, 21, 22, 26, 29, 31, 34, 71
Horizon astronomy, 59, 63, 67, 173
Horizon deviation, 64–66
Huacas, 9, 11, 14, 17, 19, 22, 36, 92, 95
Huanca, 8
Huascar, 12
Huayna Capac, 9, 11, 12, 14, 28, 85
Hurin, 21

Inca Empire, 5, 13, 14, 36
Inca Yupanqui, 8, 10, 11, 70
Intercardinal axes, 76
Inti Raymi, 23, 24, 71
Irrigation, 27
Isla del Luna, 71
Isla del Sol, 71

June solstice, 76, 86, 90, 157, 173

Kay Pacha, 18
Kenko Grande, 79–81, 173

Lacco, 82–84, 173
Lake Titicaca, 8, 18
Lau, G., 20
Lee, V., 8, 11, 13, 22

Light and shadow, 63, 70
Llactapata, 79, 90, 92, 94, 173
Lupaca, 8

Machu Picchu, 27, 66, 76, 79, 86–92, 173
Maize, 26, 62, 63, 78
Mama Ocllo, 12
Manco, 8, 18, 28
Manco Capac, 8, 28
Margolies, L., 21, 29, 31, 32, 36
Masonry, 11, 29, 30
Milky Way, 19, 20, 69, 75–77, 93, 95, 157–159
Moche, 7
Moieties, 21
Money, 22
Moon, 16, 18, 26, 37, 38, 60, 71, 78, 82, 83, 92, 93
Mummies, 12, 21, 22, 32, 71
Mythology, 8, 15

Nadir, 62, 157
Nasca, 8, 70
Niches, 32, 79
Niles, S., 9, 11, 12, 21–23, 27, 32

Offerings, 17
Ollantaytambo, 27
Orlove, B.S., 78

Pachacuti, 8–10, 14, 28
Panacas, 21, 22, 28, 79
Patallacta, 140
Paternosto, C., 9, 11, 18, 20, 23, 30, 36

Payan, 22
Pilgrimage, 9, 17, 18, 27–28, 70
Pillars, 26, 38, 61, 63, 65, 67, 73, 85, 173
Pizarro, 13
Planets, 78
Platforms, 20
Pleiades, 11, 26, 69, 72, 78–79
Polo de Ondegardo, J., 20, 86
Polygonal, 30
Pope Alexander VI, 13
Pumas, 20, 23, 36, 38, 79, 82

Quadripartition, 75–76
Quespiwanka, 65, 85
Quipus, 23

Rainy season, 26, 34, 73, 77, 157
Ranney, E., 8, 10, 16, 18, 21, 29, 31, 71
River Intihuatana, 86, 90, 92, 173
Royal history, 22
Royal Marriage, 21
Royal Mausoleum, 79

Sacred Plaza, 91, 92, 173
Sacred Valley, 12, 66, 76, 85
Sacsahuaman, 11
Salazar, L., 10
Salomon, F., 20, 63, 158
Schreiber, K., 72, 79
Serpents, 20, 23, 36, 38, 157
Shrines, 5, 9, 14, 20, 63, 79
Smallpox, 12
Solar pillars, 85
Solstices, 11, 12, 18, 24, 26, 61–63, 67, 71, 72, 77–79, 81–83, 85–87, 89, 92, 94, 157, 159, 173

Southern Cross, 157
Spanish conquest, 22, 61
Stairs, 18, 36, 38
Stanish, C., 9, 17, 18, 28, 70
Stars, 16, 18, 38, 59, 60, 69, 77–79, 158
State worship, 17
Steps, 20, 36, 38, 79
Summer solstice, 61, 159
Sun, 8–11, 16–19, 23, 25, 26, 28, 37, 38, 59–63, 66, 70, 71, 76, 78, 79, 82–84, 92, 93, 159, 173, 175
Sun-Worship, 70–71

Tambomachay, 142
Tello, J.C., 20
Temple of the Sun, 70
Terraces, 27
Textiles, 23
Tipon, 27
Tiwanaku, 8, 9
Tomebamba, 12
Topa Inca, 9, 11, 12, 14, 28

Uchu Pacha, 18
Urioste, G., 20, 63, 158

Urton, G., 18–20, 26, 60, 61, 70, 75–77, 157–159

Van de Guchte, M.J.D., 36
Venus, 78
Vilcanota, 19
Viracoca, 70
Viracoca Inca, 70
Viracocha, 8, 18
Viracocha Inca, 8

Wari, 7, 8
Winter, solstice, 61
Worship, 16, 17, 21, 70, 79, 92
Writing, 22

Yancas, 63
Yarivilca, 8

Zenith, 62, 75, 84, 157
Zenith Sun, 62–63
Zuidema, R.T., 8, 21–23, 62, 79